THE
FIGHTING
MAN

THE FIGHTING MAN

From Alexander the Great's army to the present day

Brigadier Peter Young DSO MC

Introduction by
Lieutenant General Haim Laskov, Israeli Army

The Rutledge Press
New York, New York

Facing title page *French dragoons returning victorious
at the battle of Austerlitz. The standard bearer is holding
aloft a captured Russian flag.*
Page six *One of the 'leathernecks': a US marine, still
indomitable after the hard fighting on Okinawa.*

Published by The Rutledge Press, 112 Madison Avenue,
New York, New York 10016

Distributed by W.H. Smith Publishers Inc.,
112 Madison Avenue, New York, New York 10016

First printing 1981.
Printed in Hong Kong

Library of Congress Cataloging in Publication Data
Young, Peter.
    The Fighting Man.

    Includes index.
    1. Soldiers.   2. Military Art and Science.
3. Military History.   I. Title.
U27.668   355.3'09   81-8714
ISBN 0-8317-4503-7   AACR2

CONTENTS

INTRODUCTION

Lieutenant General Haim Laskov, Israeli Army

No nation can ignore the relationship between military strength and international politics. Indeed, nation states often owe their very survival to the existence of their armed forces. Nuclear weapons may have made general conflict on the scale of the two world wars less likely, and nuclear war itself may be unthinkable as long as nuclear parity is maintained, but conventional wars on a limited scale remain with us and constantly threaten the security of individual states and world peace. In these dangerous circumstances, the problem is perhaps as much one of an anti-military bias as one of military aggression. The world situation demands unequivocally that the nations of the free world maintain substantial military capabilities for the foreseeable future. Armed forces must be recruited and paid for if a 'management of violence' is to be achieved. Fighting men must be found to man the military machine – men who are capable of independent thought and know that they will fight should the need arise. It is with the history of such men that this book is concerned. And in studying their motivations, how they lived, the factors that contributed to their successes and failures as soldiers, we come closer to an understanding of the needs and potential of the fighting man today.

The outward forms of war may have changed a great deal, but the fighting man is still an essential ingredient. The factors which make up the character of a soldier are identified in this book – discipline, loyalty, leadership, toughness, stamina, initiative and bravery. Yet although these are always present, their combination differs between different armies and in different periods, depending on the society from which they come. Reading these well-selected and instructive chapters, one must resist the temptation to seek a secret formula for military success that would hold for the fighting man in even this computerized age. It would be wrong indeed to attempt to quantify the soldier's attributes and place them in order of importance. Even if the outstanding fighting man was thus defined, the extreme circumstances in which he might find himself could not be accounted for. He may be exhausted, disillusioned or frustrated. How will he then behave? The qualities of the fighting man must remain an open-ended subject. All we can say is, 'This is how it was done at a certain place and time' – implying the essential relationship between the soldier

and the military machine in which he serves.

In painting this picture, Brigadier Young's canvas is vast. The fighting men described fought in many different kinds of war – wars of imperialism, wars of conquest, wars to assert law and order, civil wars, revolutionary wars or defensive wars. Each of these has included very different methods of warfare, from guerrilla tactics to massive concentrations of armoured vehicles. Thousands of years of military history have also witnessed changes from the use of simple infantry and cavalry formations to complex combinations of land, sea and air power. Yet, through all these changes, the military virtues outlined above have remained constant.

A military force is more than the sum of its parts because it is made up of the men that serve it. It is absurd to praise the military machine but criticize the quality of its soldiers. The strength of an army lies in its unity and, conversely, it is in the stability offered by discipline and order that the fighting man finds his individual strength. He learns confidence by subordinating his personal interest to the cohesion and purpose of his unit. It is here that the life of the fighting man may seem so different to that of people in other walks of life. But every profession has its appropriate forms of behaviour. Civilians interact among themselves to advance personal interests within the framework of law. The military aim of success in battle requires very different forms of behaviour, based on discipline and loyalty. Thus, the basic relationship described in the following chapters is the complex one between soldiers, army and society. This obviously works at various levels – from the contrast between the warrior societies of the Vikings and the feudal knights of the late middle ages to the contrast between the professionals and conscripts of the British and French armies in 1914. And while no hard and fast rules can be established, certain lessons may be drawn. Complete militarism should be avoided, with its lack of imagination or flexibility, its tendency to disregard and even disobey elected governments, but the introduction of civilian standards into military affairs may be an equal danger. The result can be an inefficient and amateurish army, at worst an armed mob. Particularly in forces recruited on the basis of compulsory national service, there will always be tension between military and civilian forms of conduct. It is better to live with that tension, accepting the criticism that the military life is harsh and demanding, than it is to sacrifice the strength to be found in military standards. Some of the best soldiers described in this book served in the armies of the American Civil War in which the tensions between an egalitarian, libertarian society and the military life ran particularly high.

Raw recruits through the ages have always brought their essential characteristics with them. It is these many and various qualities which provide a major element of the equation which forms the fighting man. Those from farming communities, for instance, often have physical toughness and a respect for the authority of their traditional leaders – whether king, tribal head man, feudal lord or commissioned officer. Alexander's army was based upon such relationships, as were the best forces of the Byzantines. Recruits may also bring with them a readiness to accommodate individual desires to the interests of group cohesion and purpose. The Ironsides of Oliver Cromwell, for example, were thinking men prepared to subordinate themselves to the effective functioning of the military machine. Then again, the recruit may be motivated by strong political or ideological feelings – be they defence of the homeland as in the early Swiss armies or the revolutionary zeal of many of the French armies of 1793.

Such enthusiasm is not always the case; most armies have either attracted personnel through the offer of attractive terms (such as in the Roman Army) or by decree – compulsory service of some kind. And, of course, there have always been societies in which war and a warrior caste were automatic. Finally, there are those who join because they enjoy the military life itself; who are attracted by life inside the machine. For if the nature of the recruit is one major element in the equation, then what the army does to those raw qualities is the other aspect – and it is from the interaction of these two forces that the fighting man emerges. The military machine must develop teamwork, and yet inspire *élan*; it must give its soldiers a mastery of technical skills, from drill and the handling of simple weapons to the utilization of sophisticated technology.

Facing death himself, a soldier does not fight merely because he is paid. He is party to a special contract in which terms like duty, discipline, country and honour play a part. It is the challenge of military leadership to develop this further through military training. Training of course, includes a great deal more – physical fitness, weapon handling, manoeuvres in the field, defensive tactics or techniques for fighting in close formations. As far as this training is concerned, it is of great interest to note the various levels of instruction: from the medieval knight who fought purely as an individual to the soldiers of Frederick the Great who were trained to have no individual fighting capacity, but to act purely in concert. The most effective soldiers, however, have both attributes. The soldier has to learn to perform his skills when under fire and dispersed from direct command. He has to know his precise role in his unit and the importance of finding within himself the strength for extra effort needed in a crisis; and at this level, training is very closely linked to the question of discipline.

The nature of discipline has not changed for hundreds of years, but its relation to other qualities has. Such factors as loyalty, initiative, responsibility and duty are increasingly stressed, so that the ability to inspire them is now the essence of leadership. The relationship of punishment to discipline has been frequently misunderstood. Where there is no discipline, punishment will not fill the gap. Some armies – such as those of the eighteenth century – seemed to create an effective machine through brutal punishments, but in fact they were shown to be inferior to the armies which came out of the French Revolution, where a different military spirit operated. The form punishment takes has also been modified with its changing role, from corporal punishment to disgrace, fines and loss of freedom or privileges. Punishment must be interwoven with sound training and fighting spirit to form a single strong rope. In some armies, punishment has indeed achieved a rigid subservience to discipline; but in others the scope for initiative and bravery has been widened by other approaches.

Leadership is, in the final analysis, a key factor. Discipline is most soundly based on a commander's ability to create and maintain a high and steady level of morale, founded on mutual trust, confidence and respect. With such leadership there is only one centre of loyalty. The commander harmonizes means with ends. He orchestrates military skills and fire power to achieve his object, leading alert, independent soldiers rather than drilled automatons waiting for their orders. Such a commander is motivated by the will to win rather than the fear of failure. He seizes the opportunity in every difficulty and ignores the difficulties found in every opportunity. Thus, the role of the great commander can often not be disentangled from the achievements of his men: from Alexander, through leaders such as Oliver Cromwell and Napoleon.

The role of leadership is wider than that of training or discipline, for the strengths of any body of fighting men are dependent for their effectiveness on the constant impact of leadership at all levels. The relationship between officers and NCOs and the individual soldier is critical to an army. In the range of its possibilities – from the Prussian officer caste of the eighteenth century to the experienced centurions promoted from within the ranks of the Roman army and from the hardened professionalism of the German officers of the Second World War to the almost mystical codes of conduct of the Japanese officer corps of the same period – lie many of the most important elements in the nature of the fighting man. Without leadership, there may be many an individual hero, but there will be no effective fighting machine.

The world has come a long way even since the last battle described in this book. New methods of waging war are being made available as the scientists introduce

8

radical developments in equipment and weapons. To be effective in action, these depend on the demands of battle – the needs of the fighting man in his attempt to gain an advantage over the enemy. Or is it true, perhaps, that the new weapons are so significant that they will come to dominate the field of battle, diminishing the role of the fighting man himself? It is quite likely that we shall not know their effectiveness for some time. But I believe it is safe to assume that technology will take on a greater and greater dominance in the future. What demands will this put on the fighting man? Military leaders will need to show great creativity in developing new tactics and organization. And this is not just a matter of reacting faster than before, but of reacting in ways that have never been tried in the past. Take speed of communication or the movement of forces and materials – what took hours now takes minutes. Accuracy of fire – directly on the target at almost any distance, nearly every missile a kill – is already with us. Now we can hear, smell or see over great distances, identifying friend or foe. These factors make it less necessary to have soldiers on the spot. On the contrary, it is up to the new fighting man to master the new technology, enabling him to obtain the information he needs at a distance through the equipment developed for the purpose. In the future, we can assume that any identified target can be hit, that the distance will be unimportant. Technology will dominate in the hands of those who are most creative in its use.

In practice, this may mean the paradox of the small, mobile team regaining its importance, with the infiltration tactics used so successfully by the Japanese infantryman in World War II returning to prominence – as long as new tactics and soldierly attributes are developed. The constant scrapping and replacement of routine procedures has to be accepted as a part of the new military doctrine. The battlefield is always changing; the fighting man must be prepared for this.

Behind all such discussion lies the possibility of nuclear war. A nuclear explosion on the battlefield would not just be a manifestation of stronger fire power – it would be a catastrophe in the face of which the fighting man is helpless. It is easier to prevent a nuclear war than to win it. But this prevention is the role of governments, not soldiers.

1
THE ARMY OF ALEXANDER

Between 334 and 326 BC, the army of Macedonia, a backward state on the fringe of the civilized world of classical Greece, marched more than 11,000 miles and created the largest empire the world had yet seen. It stretched from the Adriatic Sea to the Indus River, and incorporated the area of all the previous empires of the ancient near east – Egypt, Persia, Assyria and Babylonia. This new empire was built by one of the most effective armies the world has ever known, a war machine in which tactical control and co-ordination, overall organization and a rational command structure provided the scope for individual bravery and fighting ability to exert the maximum impact. There had, of course, been great armies before this one; the Assyrians spring naturally to mind. But the Macedonian forces are one of the first for which we have the detailed knowledge enabling us to examine how the soldier fitted into the framework of a successful military force. It is appropriate, therefore, that this study of the fighting man should begin with a look at the Macedonian soldier in the armies of Philip II and Alexander the Great.

When Philip inherited the kingdom of Macedonia in 359 BC it was an insignificant state situated in the northern part of present-day Greece and incorporating some of what is now Yugoslavia. Unlike the city-states to the south, Macedonia was a largely rural society, ruled by a hereditary monarchy, tempered however by its subject nobility. These were large landowners with dependent peasants who were bound to their superiors by a chain of duty which foreshadowed medieval feudalism. The monarchy was not traditionally absolute, the king was considered the 'first among equals'. The nobles were his 'companions' and technically they could elect and also depose him. They jealously guarded this independence which caused considerable tension when Alexander subsequently assumed the universal philosophy – and absolutist methods – of the Persian monarch he conquered.

The first step towards this new absolutism was begun within Macedonia itself by Philip, Alexander's father. Philip managed to crush his most dangerous landowners and brought their private armies under his own control, creating a large force and using it to expand his state's frontiers at the expense of Athens, Illyria and the tribes of the Danube region. The great city-states of classical Greece were in disarray; their former power had been shattered by internecine strife and by constant pressure from Persia from the east. Philip realized that the prejudices of the small autonomous states had to be overridden, both for his own expansionist ambitions and, more importantly in the short term, to resist the Persians effectively.

Left A Macedonian hoplite (left) together with a more lightly armed peltast.
Right A representation in mosaic of Alexander at the battle of the Issus. Alexander's achievements, obtained through the excellence of his army, dazzled subsequent generations.

By 337 BC Philip had formed the League of Corinth, an alliance of all the Greek city-states except Sparta, with himself at its head; he hoped to lead the forces of the League in a campaign against the Persian Empire to free the Greek cities of Asia Minor. There were economic as well as heroic reasons for Philip's expansionist ambitions: his campaigns in Greece had been costly, if successful, and he was in debt. Unfortunately he was assassinated in the following year and it was left to his twenty-year-old son Alexander to consolidate the power of Macedonia and strike eastwards against the Persians.

Alexander's whole upbringing had trained him as the commander of a large conquering army. However, he had inherited its basic organization from his father, who had been a protégé of the great Theban military tactician Epaminondas, whose ideas Philip put into practice. The army was made up of a core of 15,000 Macedonians, augmented by contingents from the League of Corinth, Greek mercenaries from the city-states and groups of allies from the Balkans. The army that crossed the Hellespont in the spring of 334 BC was around 50,000 strong. It says much for the sophistication of Philip's methods that this army remained essentially unchanged during the whole of Alexander's ten-year campaigns in Asia.

The backbone of Alexander's army was provided by the Companions, the cavalry which was led by the king himself and which originally consisted of the landed aristocracy. As the nobility was dispersed as governors of

A detail from the Royal sarcophagus at Sidon; Alexander (on the right) is leading his cavalry into battle. His readiness to lead his men in person was no fiction, and frequently put him in considerable danger.

the conquered territories and as generals of other units, the Companions gradually evolved into a highly professional cavalry unit whose members were taken from among the lesser nobility; they also took on the role of king's bodyguard on the field of battle – a responsibility of obvious importance.

There were 1800 Companions, divided into eight squadrons or *ilae*. Each Companion was responsible for his own arms and equipment which, although the Companions were referred to as 'heavy cavalry', was relatively light. Armour consisted of a helmet, a breastplate of metal or leather, a fringed skirt of leather, and greaves to protect the lower legs; the Companion also wore a belted tunic, a cloak and open-toed sandals. Arms consisted of a large round shield and a slender six-foot-long thrusting spear made of cornel wood with a metal head, and a short sword with either a straight or curved blade. Horses were small, and the primitive saddles had no stirrups; Companions could not, therefore, brace themselves for a charge with spears couched under the arm, but instead used a thrusting technique. In any case, the spear was usually dropped or broken fairly quickly and the Companion relied on his sword after the first contact with the enemy.

The Companions were complemented by light cavalry, recruited from lower down the social scale; they had only a helmet and a small shield for protection, and might be armed with lances requiring both hands for use, in which case the horse had to be guided only with the rider's knees; alternatively, the light cavalryman might use the bow or throwing spear. There were about 900 of these men in Alexander's army, and they were used for reconnaissance and skirmishing on the flanks of the main force.

The infantry were also divided into two principal types, of which the more prestigious were the *hypaspists* – the 3000 picked men of the king's personal bodyguard (although, as we have seen, this role was taken over in practice by the Companions as the king fought on horseback). Their defensive armour was similar to that of the cavalry, although the shield might be smaller and made only of wickerwork; their main weapon, however, was the *sarissa*, a pike about eight or ten feet long with a one-foot-long iron blade. For the bulk of the infantry, the *pezetaeri*, the pike was even longer – at least thirteen feet and sometimes as much as eighteen. The *pezetaeri* formed the main strength of the Macedonian phalanxes, and consisted of the landowners' peasant levies; Philip gave them the title of Foot Companions to give them a self-conscious identity to help win their allegiance to him instead of to their own local leaders.

Training in the use of the *sarissa* was of great importance for both types of infantry in the Macedonian army. The soldier had to learn to grasp his pike at a point between three and six feet from the butt, which had a metal spike to help fix it into the ground in the event of a frontal charge by the enemy. Normally, the *sarissa* would be held horizontally when moving forward to the attack, and its length caused it to wave somewhat; since the pikes of four or five ranks of men would all project beyond the front line, this must have been an intimidating sight for any foe. To increase the psycological effect, the men normally shouted the ancient war-cry '*Alalalalai!*' as they broke into a run when nearing the enemy. But the soldiers also had to learn the more prosaic manoeuvres such as wheeling on the march, advancing at an angle, and forming columns, squares and wedges, always keeping their dressing at about a yard between each file – unlike the earlier Greek armies, where the men had fought shoulder-to-shoulder. It is not surprising that the leaders of the individual files were all highly-paid veterans, the equivalent of NCOs in armies of the modern period. The infantry, too, had their skirmishers, armed with javelins, bows, and slings, although the sling could only be used effectively by someone who had practised it for years. When they were not fighting, these light infantry acted as foragers and servants for the *hypaspists* and *pezetaeri*.

The units of Alexander's army were designed to work in close co-operation, using the traditional Greek phalanx as the pivot of the attack. But the Macedonian phalanx was not identical to this original model. Use of the long *sarissa* meant that the phalanx could keep the

The origins of the Macedonian phalanx lay in the close packed infantry of the Greek city states – the hoplites, especially those of Epaminondas, the Theban general.

enemy at a distance; at the same time its very length made it difficult to handle, hence the emphasis on drill in training. The phalanx would attempt to engage the enemy; meanwhile the cavalry would charge at the right moment in wedge-shaped formation, point foremost, to smash a gap in the line and expand into it. Light infantry were deployed to prevent any attempted outflanking by the enemy and to hold off any attack until the exact conditions suitable for committing the Companions were reached.

This well-drilled army which marched through the great Persian empire was held together by various bonds. It had a feudal base, and, therefore, a natural acceptance of hierarchy and obligation; and, as we have seen, Philip and Alexander successfully integrated these elements into the discipline necessary for military effectiveness, creating a military society in which toughness and obedience were the prime virtues, and in which corporate spirit was reinforced by the position of the army as a Greek force in Asia.

The army was accompanied on the march by a surprising number of non-combatants, including physicians, poets, secretaries, surveyors, engineers, tutors, bodyguards, soothsayers, prostitutes and even philosophers, as well as the inevitable sutlers and traders. There were relatively few pack animals in the army, since the servants and the men themselves did most of the carrying. Alexander ordered that there should not be more than one servant for every ten infantrymen and one for every mounted warrior, since the soldiers carried most of their own equipment on their backs while on the march. Similarly, wagons were dispensed with in the interests of flexibility and although this meant more hard toil for the infantry, the advantages of surprise and rapid deployment for action were often used by Alexander.

The Macedonian Phalanx

The Macedonian army was built around the phalanx which consisted of a solid line of *hoplites*, normally 16 ranks deep. The smallest tactical unit in the phalanx was the *syntagma* of 256 men (illustrated below), four of which made up a regiment or *chiliarchia*. Similarly, four *chiliarchiae* formed a 'simple phalanx', supported by light infantry (*psiloi*) and cavalry (illustrated right). The phalanx was a cumbersome, though fearsome, formation and to be effective demanded the close co-ordination of the well-trained *hoplites*, cavalry, and light infantry; but also essential were the *peltasts*, equipped with a short pike, who played a crucial role as the mobile reserve behind the unwieldy close packed phalanx itself.

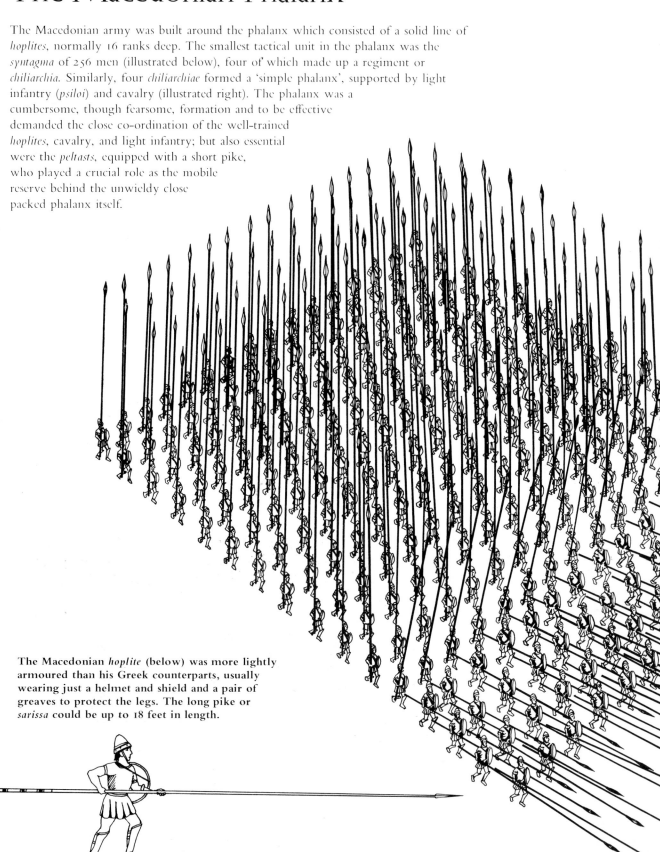

The Macedonian *hoplite* (below) was more lightly armoured than his Greek counterparts, usually wearing just a helmet and shield and a pair of greaves to protect the legs. The long pike or *sarissa* could be up to 18 feet in length.

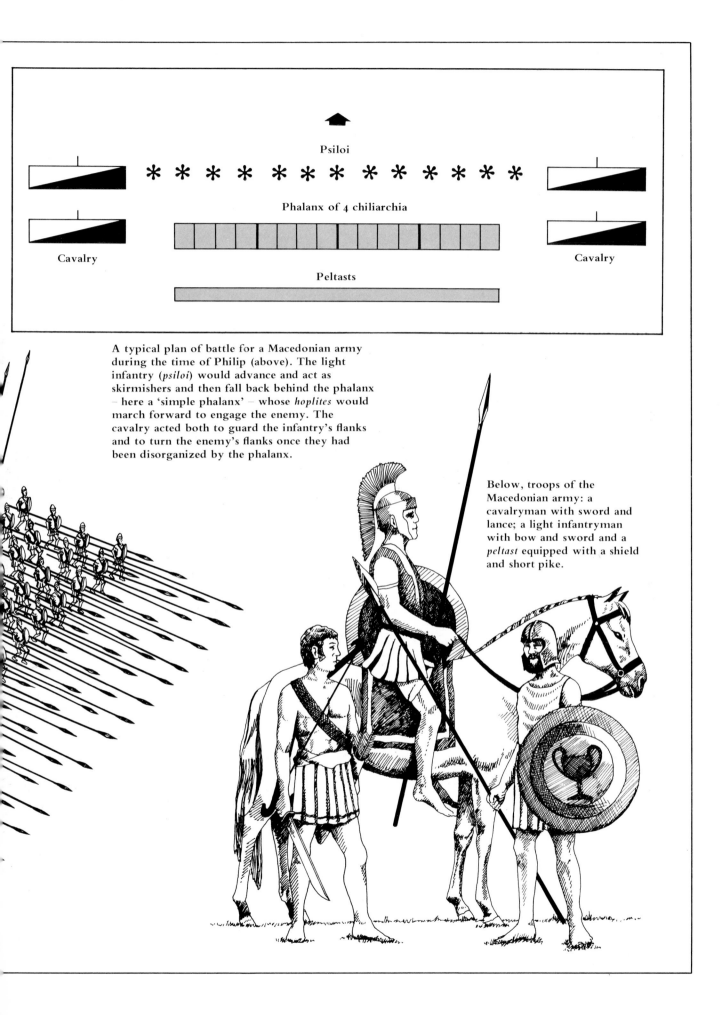

Psiloi

Phalanx of 4 chiliarchia

Cavalry

Cavalry

Peltasts

A typical plan of battle for a Macedonian army
during the time of Philip (above). The light
infantry (*psiloi*) would advance and act as
skirmishers and then fall back behind the phalanx
– here a 'simple phalanx' – whose *hoplites* would
march forward to engage the enemy. The
cavalry acted both to guard the infantry's flanks
and to turn the enemy's flanks once they had
been disorganized by the phalanx.

Below, troops of the
Macedonian army: a
cavalryman with sword and
lance; a light infantryman
with bow and sword and a
peltast equipped with a shield
and short pike.

It is now, of course, extremely difficult for us to imagine how it felt to be a soldier in Alexander's army; many of the procedures and relationships common to modern armies were clearly present, but some were startlingly different. As far as sexual relationships were concerned, for example, the men were not allowed initially to take their wives with them, and marriage to local women was not permitted until 330 BC. This strict regulation of female company is a not uncommon feature of the military life. Rather less common was the open practice of homosexuality. Homosexual liaisons were generally quite short-lived (Alexander himself had a celebrated lasting relationship with Hephaistos) and it is interesting to speculate on the effect of this overt acceptance of these attachments on the discipline of the army. The phalanx had to be an immovable block, in which every man held his place, for once the wall of pikes was breached, the *sarissa* became a cumbersome liability; and the cavalry too fought in formations where discipline was all. Certainly, the effects of these homosexual relationships cannot have been adverse, for the troops maintained their cohesion in battle in all circumstances.

The soldiers of Alexander's army seem to have been motivated by greed as much as those of other armies; booty was certainly one of their objectives, especially in the rich but fragile Persian empire. They also had a deep-seated contempt and hatred for the Persians and the people they conquered. There was particular resentment at Alexander's increasing adoption of Persian ways, which came to a head at a confrontation in Opis in 324 BC. A major grievance which the Macedonian troops expressed at this meeting was the treatment of foreign troops, some of whom had been brigaded in the Companion Cavalry, and armed with Macedonian weapons. To the veterans of the army, such equality of treatment struck at the very foundations of their exclusive *esprit de corps*.

Discipline in the army was harsh, and everyday life could be extremely uncomfortable. Rations normally consisted of bread and olives, and whatever could be foraged on the march. Even officers might be forced to march thirty miles in the height of summer with a month's supplies on their backs. Philip once castigated a soldier for washing in warm water, and he posted light cavalrymen behind battle-lines to kill any deserters.

It was with this highly disciplined and organized force that Alexander swept across Asia Minor in a series of unprecedented victories. The army easily overcame a Persian force of about 40,000 men – half of them Greek mercenaries – at the River Granicus in western Asia Minor in May, 333 BC, and they then proceeded to liberate the Persian-controlled Greek cities of the coast, moving south and east as far as the corner of the Mediterranean near the border with present-day Syria. There, in October, the main Persian army under Darius III, 100,000 strong, finally caught up with them and prepared to cut off their line of communication at Issus; Alexander was now forced to give battle against an army which outnumbered his available forces by about three

Darius, the leader of the Persian hosts which Alexander defeated during his expedition into Asia. The Macedonian military machine of 334 BC was virtually invincible.

to one. But discipline and co-ordination gave victory.

Alexander drew up his army in his favourite order of battle at an oblique angle to the enemy, with himself and his Companion cavalry on the right, the *hypaspists* to his left and slightly to the rear, the *pezetaeri* farther left and farther back, and finally the Thessalian cavalry on the extreme left rear to protect the flank of the infantry from attack by the Persian horse. After dealing with a Persian advance guard, Alexander gave the signal to attack and led his Companions and *hypaspists* in a wedge-shaped formation across the river dividing the two armies, smashing into the Persian left and scattering it.

In the centre, the Persians had counter-attacked against the main phalanx of the *pezetaeri* while they were crossing the river, and Alexander's infantry were momentarily thrust back in confusion until he turned against the flank of the Persian centre with his cavalry and *hypaspists*, taking the pressure off the *pezetaeri* and giving them time to recover and move into the attack once again. On the extreme left, meanwhile, the Thessalians held their own against the assaults of the Persian horse, destroying Darius's last hope of turning the tide; with his centre crumbling fast, he fled the field, joined by thousands of his panic-stricken men. Alexander's men pursued them relentlessly, and by the end of the day

about half the Persian army – 50,000 men – are said to have lain dead on the plain; the captives included the entire family of Darius and huge quantities of loot. Macedonian casualties totalled 450 dead, and Alexander had them buried with full military honours:

Despite his wound, he went round all the other wounded and talked to them; he collected the dead and buried them magnificently with all the army arrayed in their full battle-finery; he had a word of congratulation for all whom he himself had seen distinguishing themselves particularly bravely or whose valour he heard from agreed reports; with extra presents of money he honoured them all according to their deserts.

The Macedonian army, tested in battle and bound firmly to its brilliant young commander, was now a superb fighting force. At Arbela in October 331 it defeated a Persian force said to be twice as large as the one at Issus; again, a cavalry spearhead smashed through a gap in the Persian line 'like a flock of cranes that are flying in formation' behind the king himself; a Persian attempt to outflank the infantry in the centre was foiled by the light cavalry, positioned behind the main line for just such a purpose. In 329 the army moved into central Asia against the Scythian barbarian tribes, and here, too, the enemy usually panicked in the face of the tightly disciplined veterans of the Companion cavalry, charging at the gallop with their metal-tipped lances. In 327 they advanced through the Khyber Pass into India, crossing a

river by night and outflanking an Indian force, then successfully harassing the enemy's elephants and causing them to stampede. This victory at the Hydaspes River was followed by a march still farther east towards the Ganges, but here the men finally declared that they could do no more; they had been away from home for eight years, sometimes starving, soaked by monsoon rains, with clothing and rations rotting in the damp and heat and with weapons and buckles rusting. Alexander was reluctantly forced to turn back towards home.

As they returned westwards through the Gedrosian desert along the shore of the Arabian Sea, the soldiers must have felt that it would have been preferable to continue into India:

In places, the sand dunes were so high that one had to climb steeply up and down, quite apart from the difficulty of lifting one's legs out of the pit-like depths of the sand; when camp was pitched it was kept often as much as a mile and a half away from their watering places, to save men from plunging in to satisfy their thirst. Many would throw themselves in, still wearing their armour, and drink like fish underwater; then, as they swelled they would float up to the surface, having breathed their last, and they would foul the small expanse of available water.

Those who survived the march through the desert knew that they would eventually see their homes once again, having carved out the greatest empire in history. The struggle finished, they celebrated their victory with joy:

There was not a shield or a *sarissa* to be seen, but all along the journey the soldiers kept drawing wine in cups and drinking horns and ornate bowls; as they walked or drove, they pledged a toast to one another. Pipes and flutes and stringed instruments played loudly and filled the countryside with their music; women raised the cry in honour of Dionysus and followed the procession in a rout, as if their god was escorting them on their way.

Alexander died in the following year at the age of 33, and the magnificent army which he had led was quickly destroyed as his successors warred against each other in order to carve out their own kingdoms. Admittedly there had been cracks below the flawless exterior of Alexander's army; there was a very real distaste for his attempts to introduce Asian customs as a way of cementing together his unwieldy empire and the Macedonian nobility particularly resented the education and training of Persians in the Macedonian arts of war as well as Alexander's increasingly autocratic rule. However, in the final mutiny at Opis, Alexander called the leaders' bluff, threatening to abandon the Macedonians in favour of his new Persian favourites, which led to a full reconciliation. Under Alexander's inspired leadership, the army of Macedonia had shown what could be done with men who were properly and regularly trained to work together for the success of the army as a whole, at the expense of personal glory of the Homeric type; rather than a band of warriors it was a true army of disciplined and professional soldiers.

2
THE ROMAN LEGIONS

The achievements of Alexander's army were extraordinary, but they were dependent on the charismatic leadership of the youthful Macedonian king, whose empire broke up in the internecine strife that followed his death. With the Romans, we are dealing with an army of rigid structure, in which every individual fighting man had a set place. In the second century AD, the period we are mainly covering, men joined the army partly because they wanted to be members of just such a definite world, where everything was fixed and organized and individual initiative worked within narrow limits, and the future of the army was quite predictable.

The history of the Roman army covers some six centuries, so the social background of its fighting men was constantly changing and developing. Originally there were the citizen-farmers of the early republic who were levied in emergency for defensive purposes. But as the empire's boundaries spread beyond Italy, these troops were obviously inadequate, and around 100 BC, Marius began recruiting volunteers on a semi-professional basis. These armies overthrew the republic and their needs and wishes had to be taken into account by successive emperors. The legions could destroy them and raise new ones; hence Augustus's reforms, which brought in the professional legionary on a twenty-year contract, sent to conquer barbarians and impose the *Pax Romana* on all the Roman provinces and succeeded at least in part in curbing the political activities of the army. Later the legions were backed by local auxiliaries: for instance, the half-Romanized barbarians who signed up along the frontiers during the latter days of the empire. It could be argued that at no time was there a 'typical' Roman army, but if there was, it was during the mid-second century AD when relations between soldiers and their imperial commanders-in-chief were most stable, discipline and skill were high, and the duties of the army as fighters, engineers and administrators were most clearly defined. Here, too, the evidence is clearest, but as

with all ancient history it is sometimes necessary to extrapolate from evidence both from earlier and later dates.

During the second century AD the Roman army consisted of three main elements: the praetorian guard, the legions, and the auxiliaries. The legions were the most important, as they made up about half the army and provided its backbone in peace as well as war. Each legion, and there were usually twenty-eight or thirty of them, numbered approximately 5500 men, uniformly organized and equipped; every recruit being a Roman citizen. The emperor and people of Rome therefore expected more of him than they did of the non-citizen provincials who made up the auxiliaries; indeed the latter

Left A centurion and legionary of the Praetorian Guard. Although rather more ornately dressed than the average soldier, the equipment was essentially the same.
Right Legionaries of the second century, scuplted in relief on Trajan's column.

were raised mainly to support the legions by providing skirmishers, archers, extra cavalry, and other handy aids for the men who did the 'real' fighting.

Roman citizenship did not, of course, mean that recruits for the legions actually lived in the capital; citizenship extended throughout the empire. In the days of Augustus most of the recruits had come from the citizen populations of northern Italy, southern Gaul and southern Spain, but the greatly increased army of Trajan had brought about what was considered to be a lowering of standards; locals from the provinces, especially in the east, were drafted into the legions. Even so, they were given Roman citizenship on joining the army, (unlike the auxiliaries, who only became citizens on finishing

Essential adjuncts to the Legion: auxiliary cavalry, here shown capturing barbarians during the Dacian campaign. The scale armour is noteworthy.

their service), and the officers were still Italians. These provincials added another 100,000 men to the army, bringing the total up to 400,000 during the Dacian campaign of 105–6 AD. Aside from occasional mass conscription, however, there were plenty of volunteers for the legions. These included the sons of legionaries, who had never known any other life; the sons of auxiliaries, who were entitled to citizenship on joining the legions in view of their fathers' service, and, more and more often, auxiliaries themselves who were allowed to transfer. Although the more prosperous inner provinces of the empire provided less and less recruits, the army remained an attractive career for many provincials right up to the third century. The fixed wage, retirement pension (usually in the form of a land grant) and secure career were undoubtedly important in this.

The first step for a would-be recruit to a legion was to obtain a letter of recommendation from a friend or relative. One such letter, dating from the second century, reads in part as follows:

> To Julius Domitius, legionary tribune, from Aurelius Archelaus . . . greetings. I have even before recommended my friend Theon to you, and once again I beg you, Sir, to consider him in your eyes as myself. For he's just the sort of fellow you like. . . . And so I beg you to let him see you, and he can tell you everything. . . . Hold this letter before your eyes, Sir, and imagine that I'm talking with you.
> Goodbye.

The young man, armed with this letter, next presented himself for an interview (*probatio*) before a committee established for this purpose by the governor of the province. At this interview he had to prove that he was qualified by his background to serve in the legion, and that he did not belong to an undesirable trade. Smiths, blacksmiths, wagon-makers, butchers and huntsmen were quite acceptable, but those who worked at jobs 'appropriate to the women's quarters' were banned. These included weavers, fishermen, fowlers, cooks, bakers, and those 'brought from an inn or from employment in a house of ill fame'. He also had to pass a physical examination to ensure that he was at least 1.7m (5 feet 8 inches) tall (reduced to 1.6m – 5 feet 5 inches – by the fourth century).

Having passed the *probatio*, the new recruit was required to take the military oath, swearing allegiance to the state or, in the case of provincial legions and auxiliaries, to the emperor himself, and only indirectly to the state. Recruits also had to swear obedience to the commands of his officers. (In the days of the republic the oath had been to the officers only; it was Augustus, worried about possible *coups d'état*, who made them swear to defend the system.) If there were several recruits taking the oath, the military tribunes would save time by having one man repeat the oath, while the others merely assented by saying '*idem in me*'. The oath was taken very seriously and was renewed at the beginning of each year. If, after taking it, the soldier disobeyed orders or deserted, his commander could have him put to death without trial.

The new recruit would then be posted to his unit. Before leaving the recruiting station, he was given the substantial sum of three gold pieces as *viaticum* (travelling money). In fact, since his expenses in reaching his unit were few, this was really an enlistment bounty. Then, under the command of an officer bearing a list of their names and distinguishing marks (if any), the recruit and his new comrades marched away to join their unit,

where their names were entered on the roll with age, father's name, place of origin, height, and the name of the officer before whom they had appeared for the *probatio*.

Thus far the recruit's new life had not been unduly hard, but now came his intensive basic training. This was a thoroughly practical course, comprising marching, physical conditioning, swimming and weapon-training. Great emphasis was placed on the first of these, as the fourth-century historian Vegetius tells us:

> At the beginning of their training the recruits must be taught the military pace. For there is no point which must be watched more carefully on the march or in the field than the preservation of their marching ranks by all the men. This can only be achieved if by continuous practice they learn to march quickly and in time. For an army that is split and disarranged by stragglers is always in most serious danger from the enemy.

Not until the eighteenth century was there to be such emphasis on precision in marching, and then only in the armies of Prussia and Britain.

There were two rates of marching: the military pace and the full pace. In the summer months, a march of twenty Roman miles had to be completed in five hours at the military pace. This, allowing for the shorter Roman mile and the longer Roman hour in summertime, is about the same as the British standard of three miles per hour, including a ten-minute halt. The full pace was more rapid: at this rate, twenty-four Roman miles had to be completed in five hours.

Physical training included running, jumping, and carrying heavy loads. In the summer, swimming was added if possible. Then came training in individual fighting, which was very thorough. Vegetius writes:

> The ancients . . . made round wickerwork shields, twice as heavy as those of service weight, and gave their recruits wooden staffs instead of swords, and again these were of double weight. With these they were made to practise at the stakes both morning and afternoon.

The Roman was trained with his weapons in much the same way as the modern soldier learns his bayonet drill against dummy and groundsack. Each recruit planted his stake firmly in the ground, with 1.8m (6 feet) projecting. He then attacked with shield and staff:

> Sometimes he aimed as against the head or the face, sometimes he threatened from the flanks, sometimes he endeavoured to strike down the knees and the legs. He gave ground, he attacked, he assaulted, and he assailed the stake with all the skill and energy required in actual fighting, just as if it were a real enemy; and in this exercise care was taken to see that the recruit did not rush forward so rashly . . . as to lay himself open to a counterstroke from any quarter. Furthermore, they learned to strike, not with the edge, but with the point. For those who strike with the edge have not only been beaten by the Romans quite easily, but they have even been laughed at.

The short Roman sword (*gladius*) was held point upwards and the legionary was taught to strike upwards into the belly, avoiding the rib-cage.

If the recruit was too idle or too feeble to reach the required standard in weapon training, his ration was changed from wheat to the cheaper and less appetizing barley. 'The wheat ration was not restored to them,' Vegetius tells us, 'until they had demonstrated by practical tests . . . that they were proficient in every branch of their military studies.' Weapon-training instructors, on the other hand, were so highly regarded that they received double rations.

The *pilum*, too, was practised at the stake. The *pilum* was a wooden throwing spear with an iron head, the point of which was tempered but the shaft of which was not. This meant that the shaft of the iron head bent on impact with the ground or an enemy's shield, making it impossible to throw back. For training, an extra-heavy version was used. 'Let them also practise hurling their missile weapons at the stakes from a distance,' wrote Vegetius, 'in order to improve their marksmanship and to strengthen their right arms.' After this came training with real weapons, first at the stake and then (with blunted points) in mock combat.

With the basic elements of soldiering behind him, the recruit could combine all the skills he had learned. This was done in route marches, loaded down with not less than 26.8kg (60 pounds) and perhaps as much as 45.4kg (100 pounds) on his back. This might have comprised the

following, as described by Josephus at the time of the Jewish revolt:

> The picked infantry who form the general's bodyguard carry a lance and a round shield, as well as a saw and a basket, a bucket and an axe, together with a leather strap, a sickle and chain, and rations for three days, so that an infantryman is little different from a beast of burden.

On occasion, it seems, the legionary was required to carry as much as seventeen days' rations, presumably when an army was advancing into a desert area. Vegetius lays down the rule that:

> The young soldier must be given frequent practice in carrying loads of up to sixty pounds, and marching along at the military pace, for on strenuous campaigns they will be faced with the necessity of carrying their rations as well as their arms. Let this not be thought difficult, if practice is given; for there is nothing which constant practice does not make easy.

Second only to marching came field engineering, for it was a standing order in the Roman army that a unit, even if it halted only for one night, should fortify its camp with ditches and palisades. Of the numerous camps in Britain, some at least are supposed to have been made during training. One at Cawthorn was probably made by men of the IX *Hispana* legion when they were stationed at York.

Battle formations, of course, were rigorously practised – line, double line, square, wedge and circle. The men were taught to keep their allotted positions in rank and file without either closing up, so that they hampered each other, or opening out so that an active enemy could break through.

Three times a month, according to the regulations issued by the emperors Augustus and Hadrian, a field-day (*ambulatura*) took place. Each legionary, in full marching order, marched 16km (10 miles) out of camp and then back again, mostly at the military pace but some of the way at the full pace. This training took place in all sorts of country, ensuring that the men would be prepared for any terrain.

His training completed, the new legionary could look forward to – in many cases – a lifetime of peace. In the mid-second century, the chances were better than even that an individual soldier would never be called upon to fight. Instead, he would spend his army career in frontier garrisons of the empire. There were fewer transfers of units from place to place in the second century than there had been in the earlier days of the empire, and the frontiers – in Britain, along the Rhine and Danube, and on the borders of the Parthian empire in the east – were relatively stable; the legions, therefore, were busily

Left Practising the combined use of sword and shield, the basic legionary skill, against a stake.
Right The remains of Roman legionary boots, of stout leather construction and with iron nails in the sole, found in London.

transforming their camps into more comfortable and permanent settlements. The turf wall with its wooden palisade gave way to impressive fortresses of smoothly dressed stone. At a legionary headquarters such as Chester, for example, the base of the XX *Valeria Victrix*, there would be a stone wall with impressive gates on each of the four sides, interspersed with turrets between them. Inside, there would be a regular grid pattern of streets between barracks, granaries, workshops, administrative buildings, and the headquarters itself. This imposing structure contained not only offices but also the legion's armoury, strongroom, and 'chapel' – the room in which the emperor's statue and the standards of the legion were kept.

In addition to fortresses, the Roman legionary was obliged to build and maintain the roads of the empire. The requirement of rapid movement of troops meant that the roads had to be wide enough for six columns to march abreast, had to be made of stone so that they would be usable in all weathers, and had to be as straight and level as possible. All this meant a great deal of work, especially at stone-breaking and the levelling of natural features. Aqueducts and canals often had to be built as well, and many of these still stand as tributes to the engineering skill of the legions, as are the defensive walls, such as Hadrian's Wall in the north of England, which marked the edge of Roman civilization.

All these engineering works obviously required specialists, and those soldiers who possessed or learned

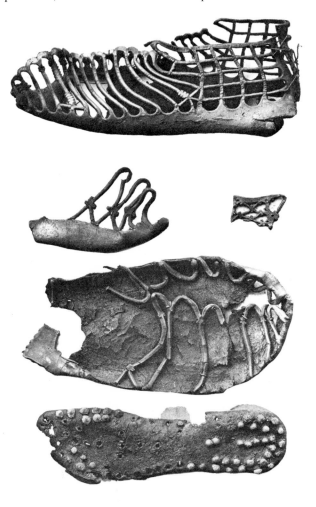

special skills were excused from many of the heavier fatigues. Among the specialists were surveyors, ditchers, smiths, wagon-makers, plumbers, stone-cutters, and lime-burners. Clerks, too, were excused. These exempted men (*immunes*) probably made up about a quarter of the men in each unit.

The *immunes*, though excused from fatigues, were still classed as ordinary soldiers as far as rank and pay were concerned. From the days of Caesar to those of Domitian (81–96 AD) the legionary received 225 *denarii* per year; Domitian increased this to 300 by paying the men four times a year instead of three. This rate remained in force until the end of the second century, and was subject to stoppages for arms and tents, clothing and food. The latter consisted mainly of wheat, eaten as porridge or bread (made into biscuits for campaigns), together with soup and vegetables. The Roman legionary was not apparently very fond of meat, although he did eat pork on occasion. Shellfish were eaten when available. The troops also demanded, and were given, wine, even though it might have had to be imported. As we have seen, barley was sometimes given instead of wheat as a punishment.

A common legionary's pay was not high but even after paying for his own keep there was usually something left for social clubs or burial societies which paid for a fine tombstone 'on an instalment plan'. However, the legionary was not left with much money to spend on riotous living. His chief pastime besides the inevitable wenching and drinking, was probably playing dice; this often took place in the public baths, where the soldier might spend his spare time relaxing with his friends. At some of the bigger garrisons there were also amphitheatres where the legionary could enjoy gladiatorial spectacles or theatrical performances. Gladiatorial displays were particularly popular; Roman society was fundamentally cruel and it is not surprising that soldiers, trained to fight and kill but rarely or never having to use their skills in practice, should take pleasure in watching others put their lives in danger. In Britain, for example, there were theatres at Isca Silurum (Caerleon upon Usk), where the II *Augusta* legion was stationed for many years, and at Corinium Dubonorum (Cirencester), where many of the veterans of the same legion were settled.

Although the legion was theoretically supposed to be self-sufficient, civilian settlements invariably grew up around a newly-established fortress to minister to the soldiers' needs and take advantage of their relatively superior spending power. Aside from the baths and the shrines, these villages might include shops selling shoes, crockery, cutlery and even souvenirs. As these settlements became larger and as more and more of the native peoples gathered in them to trade with the soldiers, public buildings eventually became necessary. Although these were in theory the responsibility of the civilian population, the skilled workmen of the legions soon began to extend their building operations outside the camp. The contact and co-operation which these activities brought about were important factors in the

23

Romanization of the conquered peoples of the empire.

The civilian settlements also housed the wives and children of the legionaries. Until the time of Septimius Severus (at the end of the second century AD), the soldier was forbidden to marry at any time during his army career, to discourage him from feeling a loyalty to a particular place; he was obliged to go wherever the service called him, at a moment's notice. But by the second century, with more and more soldiers spending their careers in fairly permanent garrison duty and with the constant need for good recruits, soldiers were encouraged to make unofficial liaisons with local women. Many of these were permanent arrangements, and in 119 AD the Emperor Hadrian decreed that their children, though illegitimate, could still inherit their fathers' estates.

The Roman soldier also had the consolations of religion. As befits a man who must (at least in theory) hazard his person in combat, he believed in placating the gods wherever he might follow the eagle standards, symbolic emblems of the legions. The eagle of the legion itself was, in fact, an object of devotion. It represented the spirit (*genius*) of the legion, and its loss in battle usually meant the disbanding of the unit. At least one occasion is known in which an official, threatened with lynching by the soldiers, grasped the eagle for sanctuary. And the anniversary of the foundation of the legion was celebrated as the *natalis aquilae* (eagle's birthday).

The gods of the Roman pantheon still had their devotees during the second century, and altars dedicated to Iuppiter Optimus Maximus (Jupiter Best and Greatest, the chief protector of Rome) or Mars have often been found outside Roman fortresses. (There were no shrines inside the fortress, except for the 'chapel' of the standards.) But the Romans also commonly worshipped the *genius* of the place where they lived, usually a barbarian god who was assimilated to one of the existing Roman deities. And then there were the personified virtues, such as *Victoria*, *Fortuna*, and even *Disciplina*; these, too, had their devotees. Before going into battle the legionary might vow to erect an altar to Mars or his local representative; in the same way an officer, before going out hunting, might placate the god Silvanus. This might not slay the boar, but it might at least preserve the huntsman's neck.

By the second century, however, there was another religion which was making great headway among the troops. This was the worship of Mithras, a Persian deity who was the personification of the Sun, and of Virtue, which implied valour as much as goodness. Like other dualistic religions, Mithraism claimed that there were two equally powerful forces constantly at war with each other; one was Light, Goodness and Spirit, the other Darkness, Evil and Material. As the heroic Mithras overcame and slew the father of material creatures, the Bull, so his followers were called upon to develop the manly virtues of strength, truth and fortitude in adversity. There was an elaborate initiation ceremony, conducted in secret, in which the devotee had to pass

Distinctive in their wolfskin uniform are these standard bearers with the eagle, the rallying point of the legion and the centre of the legionaries' loyalty.

through various perils of fire, water and blood. It was an intensely masculine cult, and its secrecy had an élitist appeal; thus it attracted a growing following among soldiers (and particularly officers) throughout the empire, far away from its original home in the east. Temples dedicated to Mithras have been found in many places where there were concentrations of soldiers, including Walbrook in the City of London.

This military life had, of course, many unattractive facets, and the soldiers were often unreliable and a danger to the peace of the empire. In peacetime, life was probably one of boring routine – a round of fatigues, guard duty, drill, physical training and route marches. The only relief would probably be when the legionary found himself working as a civil engineer. Tacitus gives an evocative description of this rather dreary life through the words of an agitator, Porcennius, active in the Rhine armies in the early years of the first century AD. Tacitus reports him as saying to his fellow soldiers:

> All these years you have been patient to the point of vice. . . . The soldier's life is hard and unprofitable. . . . It's an endless round of blows and wounds, harsh winters, summers in the field, bloody battles, or a barren peace!

Garrison life tended to soften the men, especially those

in the eastern legions. Tacitus wrote of the legions in Syria around 50 AD:

> Veterans there certainly were who had never manned an outpost nor stood a watch, who had never seen a ditch or rampart, who had no issues of helmet or breastplate. Sleek racketeers, they had done their military service in towns.

Generals with over-ambitious projects were often bitterly resented: the late-third-century emperor Probus was actually murdered by his men for keeping them employed in endless schemes for land reclamation. A legionary in Egypt in 107 AD, writing home to tell his family that he had been given a clerical job, noted: 'I'm getting on all right . . . so far I haven't been caught by any fatigues like cutting building-stones.'

Tacitus complained that the wide recruiting base gave the army no common moral code and led to indiscipline and inevitable corruption. He wrote of the army which sacked Cremona during the first century AD that it was 'diverse in language and custom, made up of Roman citizens, allies and foreigners, embraced every kind of lust, knew no common code,' and 'would stop at nothing'.

Politics, too, had its place in the Roman soldier's life. Although disobedience to orders was punishable with death, the legionaries seem to have taken part in noisy demonstrations from time to time. Their natural loyalty was always to their commander, rather than to the army or state as a whole, and in the later years of the empire the legions found they had the power to make and unmake emperors, usually their own legionary commanders, whether the candidates were willing or not. This was an inevitable development since the emperor's position was dependent on military might, initially that of his own praetorian guard. It was not until the middle of the first century that the army outside Rome discovered the extent of its power, following the death of Nero; it also showed the potential vulnerability of the empire as legions in Spain, Germany and the east put forward their own imperial candidates. Higher ranks obviously had potential goals in mind; the ordinary soldier's forays into the political arena were more probably prompted by the hope of rich prizes (*donatives*) which the imperial candidate was obliged to promise to the troops who supported his claim.

In spite of this threat it posed to the internal political stability of the empire, the Roman army of the second century was, by and large, an extremely effective fighting and peace-keeping force. The thorough organization which permeated all aspects of the military life found its true expression on campaign and in battle, where a methodical approach to all aspects of warfare reduced risk and guaranteed success.

For a major campaign, a large force would be assembled from various garrisons. Every step would be strictly planned. The Jewish historian Josephus described their order of march, for example; first there were lightly armed auxiliaries who acted as scouts, looking out for ambushes; then came detachments of the legions, both

A Roman auxiliary cavalryman, armed with round shield and spear. No stirrups were used.

infantry and cavalry, as a vanguard. Road-makers and men detailed to construct camps followed; after which the baggage train of the general officers, escorted by cavalry, was able to proceed. The commander-in-chief with his picked guard of both infantry and cavalry would follow, with the rest of the small cavalry units attached to the legions and the mules loaded with battering rams and other siege equipment. The commanders of the individual legions, the eagle-standards, trumpeters and the main body of the legionary infantry would come next, followed by their own baggage; the remainder of the auxiliary infantry, except for a detachment (with some auxiliary cavalry and legionary infantry), formed the rear-guard. In difficult circumstances the army marched in a square, with the baggage in the centre. There was also another defence for the Roman army on the march, which commanders ignored at their peril: the intended route must be kept secret.

In preparing for a battle, the choice of ground was of cardinal importance. High ground was valuable, not so much for observation as to enable the legionary to hurl his *pilum* with more force and to give impetus to his attack. On the other hand, ground covered with trees or broken up by streams and valleys would not allow the tightly organized and drilled Romans to operate effectively. As with all armies, the Romans would try to get the sun and wind at their backs so that the enemy soldier would be dazzled and would have the dust of battle in his face.

The legions would normally be drawn up in six ranks, with the veterans at the front to set an example for the untested recruits. Each man required 0.9m (3 feet) of space in the line, with 1.8m (6 feet) between the ranks. An army of 10,000 infantry would thus occupy a front of about 1.4km ($\frac{7}{8}$ of a mile). On the wings of the infantry body would be placed the cavalry, whose duties were to prevent the centre from being outflanked and to pursue and cut down the enemy once their ranks broke.

The Romans could attack by a turning movement of either the right or left wing of the infantry, the other wing holding fast, or by an advance of both wings, with the centre held back. This latter form of pincer movement was dangerous, as it could end with the front line in three separate bodies. A refinement was to screen the centre with auxiliary light infantry – archers and slingers – to keep the enemy preoccupied while the wings developed their attack. Then there was the wedge formation, similar to that used by Alexander's cavalry, used to thrust through a weak point in the enemy line and then expand to right and left. Behind the front line the Romans kept a tactical reserve, ready to move either to the right or left to plug any gaps in the line. This was in addition to the main reserve of fresh troops who could be brought into the battle at the commander's discretion.

Given firm leadership, discipline was the key to victory for the Roman soldier. In 60 AD, for example, Suetonius Paulinus, with some 10,000 legionaries of the XIV *Gemina* and XX *Valeria Victrix* plus auxiliaries, was more than a match for Queen Boudicca and her British horde, which certainly outnumbered his force by at least two or three to one. Paulinus reduced Boudicca's advantage in numbers and possession of chariots (which the Romans did not use for warfare) by posting his army on a ridge, his flanks resting on thick woodland. Boudicca's men and chariots were compelled to make repeated charges uphill, while the Romans stood their ground and unleashed a devastating volley of javelins. Each volley was followed by a charge in wedge formation with shield and sword. The Roman wall of tightly packed cylindrical shields caused enemy missiles to bounce off, while the short Roman stabbing sword was far quicker and easier to use than the long British sword. By the end of the day, it was said, the British had lost 80,000 slain, almost certainly an exaggeration, but the overwhelming nature of the Roman victory was not in doubt. The legionaries had been thoroughly trained, they fought as a unit and knew that their comrades were supporting them, and they absolutely refused to break and pursue the enemy for a short-term advantage.

It was in the heat of battle that Roman discipline and weapon training were given full scope. The legionary was not afraid to come to close quarters with his immediate opponent; indeed the best method of fighting most barbarians was to come right into them, knocking their more cumbersome weapons away with the large curved shield and thrusting upwards with the short *gladius*. And although trained as an expert individual fighter, the legionary was always aware that his prime duty was to maintain the unbroken front of the unit. The Roman legionary's morale was high; he was always conscious of his superiority to his foe as an individual fighter and confident that even in the relatively open order of the normal battle line the formation would be maintained, and he could concentrate on his own immediate task without considering the possibility of a general collapse around him. All the methodical structure of the Roman army, permeating the soldier's whole life, reinforced this confidence, which, more than any mere technical advantage, was what gave the legion the advantage over its foes for so many centuries.

After the battle, too, Roman organization was precise for those who had been wounded or had particularly distinguished themselves. For some of the wounded there was little hope: men scalded in their armour by burning oil thrown from a besieged fortress were not likely to recover. But the extraction of darts and arrows, together with simple amputation, were not beyond the skill of the legion's medical staff. Anaesthetics other than alcohol were unknown, but crude antiseptics were at least available, among them pitch, turpentine and a variety of oils. Each legionary headquarters had its hospital, staffed by trained surgeons and by sick-bay attendants from the other ranks.

Decorations were awarded with a fairly liberal hand. For the legionary there were *torques* (necklaces), *armillae* (armbands), and *phalerae* (ornamental bosses worn on the breastplate). Saving the life of a comrade was rewarded by the *corona civica*, while the first man over the wall of a besieged town might win the *corona muralis*. Some decorations were confined to certain ranks. The *hasta pura*, a silver spearshaft, for example, was awarded only to officers. In the brave days of the Republic one hero, L. Siccius Dentatus, active around 45 BC and a veteran of a hundred and twenty battles with forty-five wound scars – all in front – is reported to have received eighteen *hastae purae*, twenty-five *phalerae*, eighty-three *torques*, and so many *armillae* and *coronae* that it would be tedious to name them.

Promotion, too, was possible for the ordinary legionary. It has already been mentioned that men with specialized duties (the *immunes*) were exempted from the heavier fatigues; this was the first privilege the soldier could acquire, although it made no difference as far as rank and pay were concerned. If the soldier could read and write, however, he might move up to the grade of *principalis*, in which case his first job would be to act as orderly-room sergeant for his century (which by the time of Hadrian had been reduced to eighty men rather than one hundred). Later he might move on to the post of *signifer* or standard bearer, in which case he would also be in charge of the men's savings. These were augmented by the emperor's periodic *donatives*, and were deposited in a vault under the shrine of the legion's eagle. Or he might become an *optio* (lieutenant). The *optio* might either remain in the century as second-in-command or might be moved over to join the staff of the legionary headquarters. Staff officers often had judicial functions in the empire's provinces. It was every soldier's ambition to be promoted after good service to the rank of centurion, which was the highest post that the legionary could normally aspire to. There were sixty to the legion, and candidates for the centurionate needed the backing of the provincial governor and as many recommendations as they could get; their military records might have to be

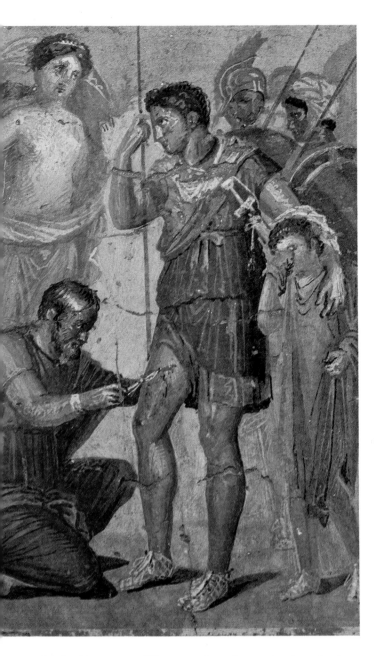

Medical treatment. The extraction of arrowheads and other such simple surgical tasks was always possible, but serious wounds were difficult to deal with.

sent for examination to Rome before they were promoted.

Once the candidate received the twisted vine-stick, which was the centurion's badge of rank, his life would change dramatically. As a *principalis* his pay might have been two or three times as much as the ordinary legionary's; every centurion received at least seventeen times the basic rate. The centurion had to pay for his own horse, handsome clothing and elaborate armour out of this, but his pay was also a just reward for the level of responsibility which was his. It is true that centurions could be very cruel to the men in their command, especially in the event of mutiny, desertion or other dereliction of duty. They were also known to augment their incomes by fining their legionaries in return for

exemption from unpopular fatigues, so it is not surprising that in the event of mutiny individual centurions were often the first victims of their men's wrath. On the other hand, the centurion did have to work hard. As the commander of a century he had to post the guards, inspect the troops, train the new recruits, and keep track of the arms and equipment (although he had two clerks to help with the latter). In battle, it was the centurion's duty to keep the men in line and set an example of bravery for his men to emulate. Later he might be called upon to join the legion's council of war or to command a cohort consisting of six centuries. Some centurions, it is true, were promoted from outside the legion – perhaps from the praetorian guard – but most of them rose through the ranks and knew their job and their men. They were dedicated, hard-working officers to whom much was given and from whom much was expected. Although some were promoted to even higher rank as military tribunes, most of the latter were young, political appointees of senatorial rank who would only stay with the legion for three or four years before moving on to another administrative post in the empire. It was the centurions who were the real professionals, providing the continuity which was essential to the army.

Promotion was for the few; what most legionaries could look forward to was retirement with a gratuity at the end of a twenty-year career. During the first century he would have been obliged to settle in a veteran's colony with other former soldiers, probably in a distant land. Verulamium (St Albans) and Corinium Dorbunnorum (Cirencester) were examples of such colonies in Britain. Luck might dictate how long he would enjoy his retirement; in Boudicca's rising, the veterans of II *Augusta* at Corinium suffered nothing more than a few extra guard duties, while the veterans of IX *Hispana* at Verulamium were massacred, the survivors and their families being impaled on sharp stakes to die in agony. The practice of sending former soldiers out to colonies in the remoter parts of the empire was an important part of the programme to bring Roman civilization to the outlying provinces. By the second century, however, such colonies were no longer being established; local recruitment for service in local garrisons was becoming the rule, and the discharged soldier would leave the camp to resettle nearby, among his compatriots.

As a soldier, the Roman did not fear death, but he dreaded the prospect of oblivion. As we have seen, many legionaries made contributions to burial clubs so that a proper tombstone could be made in commemoration of the individual's life and service. In other cases, the deceased veteran's family would erect the memorial. An example is that of Julius Valens at Caerleon: no doubt his wife Julia Secundina and son Julius Martinus ensured that his death was marked with the proper rites and the funeral feast, as well as recording his service with the II *Augusta* and his death at the ripe old age of one hundred. Julius's wish has been fulfilled, and his fine stone monument ensures that we can still recall his name and service today.

The Roman Legion

Just as the phalanx was the cornerstone of the Macedonian fighting machine, so the legion was the basis of the Roman army. The legion, however, was much more flexible than the phalanx, and was capable of forming a number of different formations to meet different situations. The legion developed over the years as Rome went from republic to empire, and changed from a citizen army to a professional force. A particular strength was the integration of the best tactical methods and weaponry of their opponents. From the Greeks, for example, they took many ideas for siege engines, and improved on them so that artillery was a basic part of any siege operation. The Romans were not merely borrowers, however, and in the conduct of sieges, they frequently used one of their own developments – the famous *testudo* (tortoise) formation, in which a group of infantrymen formed a solid block with shields covering all directions to protect themselves from enemy missiles. This formation is shown on the right.

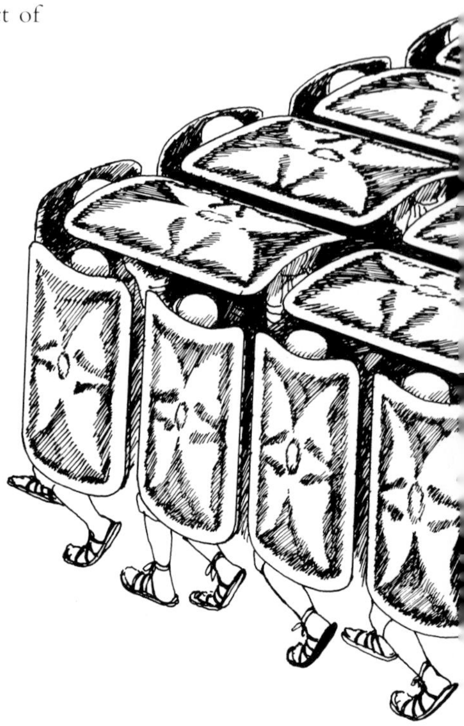

The weapons of the legionary were quite simple and standardised during the period of the empire. The sword (*gladius*) was short for stabbing, and the javelin (*pilum*) had a head which bent on impact, so that it could never be thrown back.

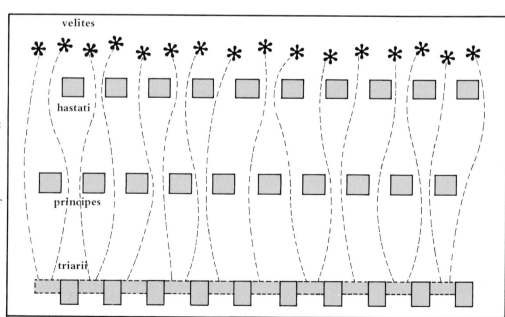

Left: the early Roman legion was composed of four distinct types of soldier. In front were the *velites*, the younger men who acted primarily as skirmishers; secondly the *hastati*, thirdly the veteran *principes* and fourthly the *triarii*, the oldest members of the legion who were reinforced by the *velites* once the battle had begun in earnest.

The basic unit was the maniple of 120 men (those of the *triarii* had only 60 men); four maniples (one each of *velites*, *hastati*, *principes* and *triarii*) formed a cohort, and ten cohorts a legion of around 5000 men.

velites

hastati

principes

triarii

The Romans used a variety of siege weapons, and two of the best known are shown here.

The catapult worked on the torsion principle: the arm of the catapult was activated by a tightly twisted skein of sinew or cord, and was winched into the firing position by a windlass. The projectile would normally be a rock.

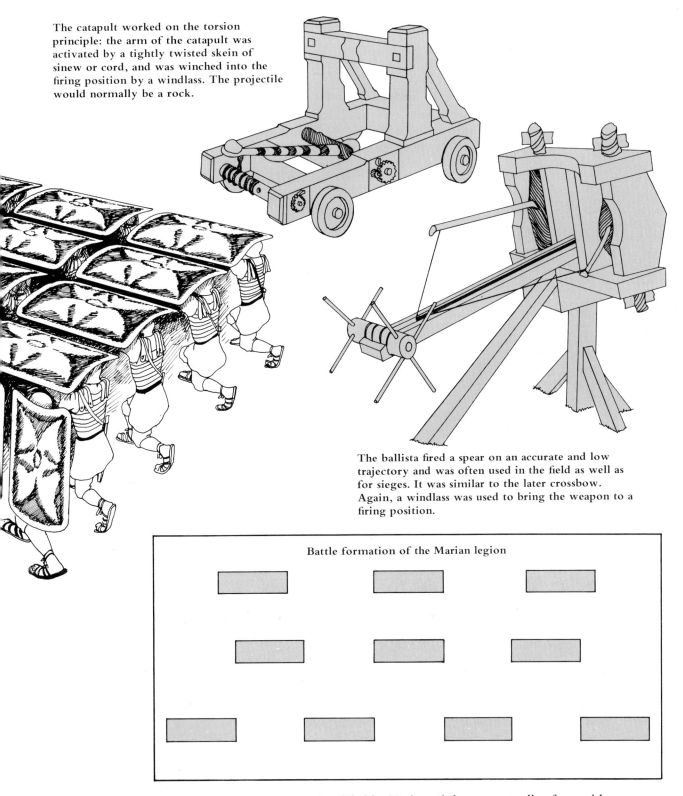

The ballista fired a spear on an accurate and low trajectory and was often used in the field as well as for sieges. It was similar to the later crossbow. Again, a windlass was used to bring the weapon to a firing position.

Battle formation of the Marian legion

The legion was simplified by Marius as it became a standing force with near-professional members at the end of the 2nd century BC. The cohort replaced the maniple as the main unit and the distinction between the various classes of soldier died away, although the names continued in use. The 10-cohort Marian legion was the basis for the imperial legion.

3
THE SOLDIERS OF BYZANTIUM

The eastern empire of Byzantium maintained one of the most efficient armies of its time. It was the direct heir to the imperial Roman army, but changing circumstances led to changing organization. The earlier emperors had generally been scared to use local forces except as auxiliaries, preferring to keep the army centralized, professional and separate from the rest of the population. Barbarian encroachments showed the weakness of this system: from the fifth century onwards, self-defence was the key issue and what were initially measures of expediency led to the formation of a remarkably flexible fighting force based on local recruitment and geared to local needs.

Indeed the army was probably the single most important factor in the survival of the Byzantine empire. Over the 1000 years of its existence the fortunes of the empire waxed and waned: in turn, the army had to fight the Ostrogoths, Visigoths, Lombards, Berbers and Vandals in the sixth century, the Persians and Arabs in the seventh, the Bulgars in the ninth and tenth, and the

Normans and Turks in the eleventh and twelfth, as well as having to contend with the constant internal strife inevitable in a large and unwieldy empire.

A rational and practical approach to military affairs was the Byzantine hallmark; and until the disastrous consequences of the defeat at Manzikert in the eleventh century deprived the empire of its main source of recruits, the well trained Byzantine fighting man was usually a match for his foe.

From the sixth century AD on, the Byzantine army's role was largely a defensive one, trying to preserve the integrity of the empire's boundaries rather than conquering new lands from the barbarians. An attack might come at any moment against some point on the long

Left *A 10th century Byzantine heavy cavalryman, wearing lamellar armour and chainmail.*
Below *This mosaic at St Sophia demonstrates the emphasis on the close links between the emperor and empress (right and left) with the Holy family (centre).*

frontier of the empire, and a concentration of forces for any purpose other than defence was risky.

In any case, the Byzantines did not regard the soldier's life as a glorious one, as some of the peoples of the West did during these centuries. The soldiers of Byzantine were not cowardly; it was simply that they considered that war, although sometimes necessary, was never desirable, and even the fight against the Moslem infidel was to be deplored; a fallen warrior was viewed more as a failure than as a hero. Prowess on the field of battle was admirable, perhaps, but not particularly desirable either. A high priority in the conduct of a campaign by a Byzantine general was the preservation of his army's resources of manpower. If the enemy could be defeated by diplomatic means – including fraud, bribery, subversion, and propaganda – so much the better. The Byzantine took a civilized attitude to war: they treated their prisoners well, stood by their treaties (once they were signed), and were generous to their enemies in defeat.

The keystone of Byzantine military success was analysis – of their own potential, their logistics, the terrain, and the enemy; their work in this area would not disgrace a modern staff college. This can be seen clearly in the surviving military manuals produced during the imperial era, including the *Strategicon* of the Emperor Maurice, written *c.* 590 AD, the *Tactica* of Leo the Wise from about 900, and a work on frontier warfare written by one of the Emperor Nicephorus Phocas's generals

The rugged countryside of the Taurus mountains, one of the main invasion routes for the Arab armies, and where the Byzantines fought some of their major battles.

about 960. These works show that although the basic principles of Byzantine warfare changed little over three centuries, objective analysis led to many adaptations in details.

The empire at first relied on locally raised troops to defend particular sections of the frontier. This early system collapsed in the seventh century when the provinces of Africa, Egypt and Syria were overrun. Thereafter, the main strength of the army lay in the *themes* or army-corps districts, the two main areas being Armenia and Anatolia. Initiated piece-meal by the emperor Justinian, and formalized by Justinian II, the *themes* combined civil and military jurisdiction under the command of the general or *strategos*, who had the title of *Patrician* and had his own extensive staff, including officials with judicial functions. Land was granted to troops in exchange for their military service. Grants were made according to length of service and type of arm: cavalrymen were allowed land worth at least four pounds of gold; waste land was given over to volunteers to exploit, sometimes even to prisoners of war. Soldier-farmers, too, would have the legal status of their smallholdings confirmed if they submitted to the duty of serving with horse and sword when called upon. They had to pay land-tax, but were freed from the burdens of roadmaking, engineering or contributions in kind.

The organization of the empire into *themes* did not happen everywhere at the same time, but grew gradually and changed as it did so. The power of the *strategoi* led to frequent attempts to seize the imperial throne, and the *themes* were therefore reduced in area over the years in order to avoid the concentration of too much military strength in the hands of a single general. At the end of the seventh century there were seven *themes*; by the year 900 there were thirty. For the same reason the emperors seem by that time to have adopted the practice of rotating the *strategoi* between the various *themes* and all the senior officers – not just the generals – were appointed personally by the emperor. They came from the best families of the Byzantine aristocracy, as Leo the Wise observed: 'Nothing prevents us from finding a sufficient supply of men of wealth and also of courage and high birth to officer our army. Their nobility enables them to win the greatest popularity among the troops with the occasional and judicious gift of small creature-comforts.'

Each of the *themes* was divided into two or three brigades or *turmae*, led by turmachs, and each of these into smaller units. The basic formation was the *numerus* or *banda*, consisting of anywhere between two and four hundred men – the numbers were deliberately kept vague and unequal so that the enemy could not calculate the size of the Byzantine army. Towns were defended by one or two units of *numen*, usually infantry, drawn from the urban population. None of the regiments were permanent or professional in the ways the legions of the old Roman army had been, except the regiments of mercenaries, known as bucellarians, that formed the imperial guard.

Recruitment and training for the local militia was undertaken by special cadres who called the inhabitants of an area to a meeting for this purpose. Each man was required to bring a pair of horses and sometimes chariots as well. Failure to obey was punishable by death. The cavalry was mainly drawn from the owners of smallholdings who had an hereditary obligation to serve. Recruits earned one *nomisma* in their first year, two in their second, rising to a maximum of twelve.

The mainstay of the army was the cataphract, a heavy cavalryman armed with both a bow (a weapon despised by their predecessors in the Roman legions), and a lance. The cataphracts were carefully selected in accordance with the principles set forth by Leo the Wise:

> The *strategos* must pick from the inhabitants of his *theme* men who are neither too young nor too old, but are robust, courageous and provided with the means so that, whenever they are in garrison or on expedition, they may be free from care as to their homes, having those left behind who may till their fields for them. And in order that the household may not suffer for the master being in service, we decree the farms of soldiers shall be free from all exactions except the land-tax. For we are determined that our comrades . . . shall never be ruined by fiscal oppression in their absence.

Men with family responsibilities, therefore could avoid military service; the widowed mother of Saint Euthymius the younger (born *c.* 820) married him off early so that with two women to support and a farm to manage he could be excused.

The cataphract was equipped with a conical helmet adorned with a horsehair crest, a mail shirt reaching to his thighs, gauntlets, and greaves or boots. A light cotton surcoat worn over the mail shirt was dyed a distinctive colour to indicate his unit, as were his helmet-crest, shield and lance-pennon. Cataphract squadrons were used in mass formations, so their warhorses might be protected with mail or scale armour especially if serving in the front line, and weapons would include the lance, a broadsword, a bow (with the quiver of arrows strapped

A relief of a Byzantine soldier carved onto a rock face in southern Asia Minor, where armed readiness was a constant necessity to preserve the integrity of the empire.

to the heavy saddle), a dagger, and often a small axe. The cataphract was thoroughly trained in the use of all weapons, with special emphasis on archery practice.

The infantry (*numeri*) were divided into two types, of whom the light infantry were mainly archers. They wore tunics reaching to their knees and large, broad-toed boots; their quivers (each containing forty arrows) and small, round bucklers were carried strapped to their backs. As many as possible wore a light mail shirt; others had to rely on scale armour made of horn or even on buff coats strengthened with steel plates. They wore no helmets, and Leo insisted that their hair must be kept short. They were subject to frequent archery practice, bearing in mind the warning of Leo the Wise: 'The neglect of this arm has caused many of the defeats suffered by the Romans.'

The heavy infantryman – the *scutatus* – took his name from the large round shield which he carried. He was armed with a lance, sword, axe (the latter with a cutting edge on one side and a spike on the other) and wore a helmet similar to that of a cataphract. Like the cataphract too, his shield, surcoat and helmet-plume were uniform in colour within his unit.

The soldier's material needs were well looked after. He was well paid, and was able to hire men or boys – either bondmen or free servants – to save himself from menial tasks. Leo stipulated that even the poorer troops should have an attendant for every four or five soldiers, as well as a pack-horse to carry extra luggage which could not be easily strapped to their own saddles. When moving through desert country the troops were entirely dependent on their attendants, who were responsible for foraging. Women camp-followers were discouraged, but other luxuries were provided: an Arab observer noted that 'Doryleum possesses warm springs of fresh water, over which the emperors have constructed vaulted buildings for bathing. There are seven basins, each of which can accomodate 1000 men. The water reaches the breast of a man of average height, and the

This cover of a tenth century casket (depicting a biblical scene) shows how Byzantine infantry were equipped and armoured during this critical period.

overflow is discharged into a small lake.' Every Byzantine army had its engineer corps, which was responsible for marking out the boundaries of the night's camp and allocating the station of each corps within it. When the main body of the army arrived, the infantrymen would begin digging the perimeter trench with the spades and picks carried in their carts, following the line of the ropes laid by the engineers. Other troops, meanwhile, would already be on picket duty. As night fell, the men would retire to their tents; for the infantry, each of these supplied shelter for sixteen men, while officers had shelters of a size and splendour befitting their ranks.

Each band of infantry or cavalry had a physician, a surgeon and between six and eight bearers (*depoutatoi*) who were responsible for removing injured men from the battlefield. The bearers carried flasks of water and had two ladders attached to the left sides of their horses to help lift the wounded onto their mounts; they received bonuses for each man rescued during the battle.

The Byzantine soldier was, then, part of a proper military structure; and the mainstay of the army, the cataphracts, were individuals of considerable social status whose linked position in civil and military life had been carefully regulated by the central government. Yet the cataphract was not a professional soldier as we understand the term. Rather, he was an extremely efficient local militiaman. The non-aggressive character of Byzantine war meant that the soldiers were usually fighting in defence of their own land, their own possessions and their own families – potent motives for a locally-based militia.

Various cultural factors were used to reinforce the army's effectiveness against the forces of Islam, or western and slav barbarism. The Orthodox faith played, for example, an important part in the life of the Byzantine soldier. Morning and evening were marked by the singing of the Trisagion (the 'thrice-holy hymn') from the Liturgy, and the password of the day was nearly always a sacred text. Battle-standards often bore the figures of warriors of the faith like St Michael, St George or St Demetrius, and soldiers were urged to make their confessions before each battle. As the troops stood in battle array waiting for the onslaught, they would be addressed by the *cantatores*, professional orators whose job it was to stir their spirits with messages such as 'We are fighting God's cause; the issue lies with him, and he will not favour the enemy because of their unbelief'. On the other hand, St Basil decreed that any soldier guilty of actually killing someone in battle should abstain from making his communion for three years.

During the period of the Iconoclasts, the strong reforming movement in the church was reflected in a strict moral code for the army. Adulterers could not serve and soldiers could be cashiered if they connived in any lapse of chaste behaviour on the part of their wives. Soldiers should have no connection with trade which might lead them to corruption or simply deflect them from their prime defensive role. Desertion was punishable by burning or crucifixion. Other punishments included flogging, or shaving of the head, beard, moustache or even eyebrows.

The adaptable, well organized Byzantine forces, bound together by a common religion and with a sound motive for fighting in defence of their own prosperity knew the character of the enemies they were fighting, and used this knowledge at every opportunity. The western barbarians were considered rash and insubordinate, and so attempts were always made to break up their formations and engage their forces piece-meal; the slavs were to be fought in open terrain wherever possible, and the Saracens, dangerous but easily demoralized, were refused battle for as long as possible, while their morale was affected by attacks on their communication.

This heyday of the Byzantine system of warfare based on the landed cataphract ended abruptly in the late eleventh century, when political intrigue contributed to the defeat of Manzikert, which was followed by Turkish occupation of central Asia Minor, the heartland of the system of *themes*. Of the remaining provinces of the empire, only Thrace contained the economic and social structure necessary for the continuation of the system, and this single province could not provide sufficient men, and especially not to defend the eastern frontier where the threat of invasion was most constant. The empire was unable to reverse its decline after Manzikert; in 1204 Constantinople was sacked by crusaders, and in 1453 it finally fell to the Ottoman Turks.

A manuscript depicting the 11th century Islamic armies. These were the troops which inflicted the crucial defeat of Manzikert upon the Byzantine forces in 1071.

The Byzantine Empire

The Byzantine empire was a major power from the fall of Rome in the 5th century AD to the sack of Constantinople by the Crusaders in 1204. The reason for this long period of success lay mainly in the effectiveness of its military institutions. The system of Themes (see map below) provided well regulated local recruiting grounds, and to buttress the local forces were the central forces based in Constantinople, known as the Tagmata.

Detailed study of tactics and strategy – to a degree unknown in the West – was an essential requirement for all Byzantine officers. The ideal was always to lure the enemy into a trap, and to annihilate him with the minimum of risk, as is shown in the diagrams right.

Although chain mail was used, the Byzantines seem to have preferred lamellar armour, consisting of small strips of iron or toughened leather laced together to form a continuous flexible sheet.

Far right is a Byzantine cataphract, wearing a mail suit under a quilted covering. The horse is covered in lamellar armour. Centre right is an archer with lamellar corslet and composite bow. Near right: the composite bow. A long strip of sinew was attached to the front of the central wooden core while two pieces of horn were glued to the back. The whole construction was usually bound with leather and had an effective range of up to 400 yards.

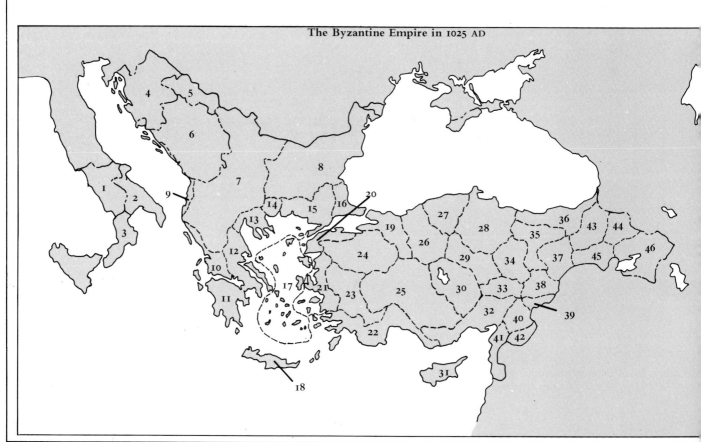

The Byzantine Empire in 1025 AD

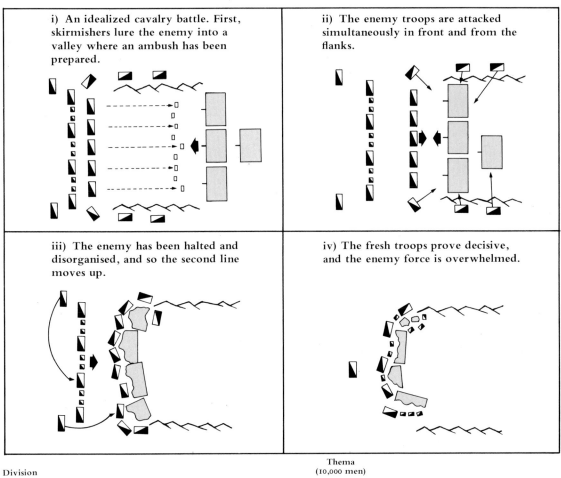

i) **An idealized cavalry battle. First, skirmishers lure the enemy into a valley where an ambush has been prepared.**

ii) **The enemy troops are attacked simultaneously in front and from the flanks.**

iii) **The enemy has been halted and disorganised, and so the second line moves up.**

iv) **The fresh troops prove decisive, and the enemy force is overwhelmed.**

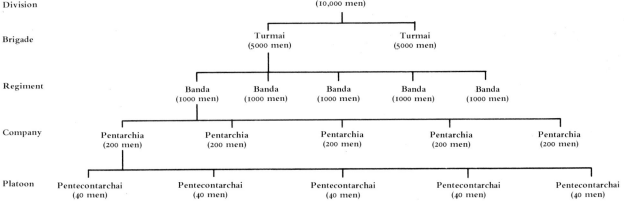

Division — Thema (10,000 men)

Brigade — Turmai (5000 men) / Turmai (5000 men)

Regiment — Banda (1000 men) / Banda (1000 men) / Banda (1000 men) / Banda (1000 men) / Banda (1000 men)

Company — Pentarchia (200 men) / Pentarchia (200 men) / Pentarchia (200 men) / Pentarchia (200 men) / Pentarchia (200 men)

Platoon — Pentecontarchai (40 men) / Pentecontarchai (40 men) / Pentecontarchai (40 men) / Pentecontarchai (40 men) / Pentecontarchai (40 men)

Both the number of Themes and the Byzantine military organisation changed during the course of the empire's development. The number of men each Theme could supply varied, and the organisational diagram is essentially a theoretical guideline.

The Thematic System

1 Lombard Principalities
2 Lombardia
3 Calabria
4 Croatia
5 Sirmium
6 Serbia
7 Bulgaria
8 Paristrion
9 Dyrrachium
10 Nicopolis
11 Peloponnese
12 Hellas
13 Thessalonica
14 Strymon
15 Macedonia
16 Thrace
17 Aegean Islands
18 Crete
19 Optimaton
20 Abydos
21 Samos
22 Cibyrraeots
23 Thracesion
24 Opsikion
25 Anatolikon
26 Bucelarion
27 Paphlagonia
28 Armeniakion
29 Charsianon
30 Cappadocia
31 Cyprus
32 Seleucia
33 Lycandus
34 Sebastea
35 Colonea
36 Chaldia
37 Mesopotamia
38 Melitene
39 Euphrates Cities
40 Teluch
41 Antioch
42 Aleppo
43 Theodosiopolis
44 Iberia
45 Taron
46 Vaspurakan

4
THE VIKINGS

The raids of the Vikings which began in the late eighth century took Christian Europe by surprise. The Northumbrian monk Alcuin wrote in 793 AD:

It is nearly 350 years that we and our fathers have inhabited this lovely land, and never before has such a terror appeared in Britain as we have now suffered from a pagan race, nor was it thought that such an inroad from the sea could be made.

It is likely that sporadic earlier raids had been made, and that Scandinavian traders would have made legitimate contacts with most of the areas that later came to fear their presence. But from this date, the Vikings began the two hundred years of constant attacks on Christian Europe for which they are traditionally famous.

Posterity's view of the Vikings has been inevitably coloured by the doleful cries of their victims:

They go by horse and foot through hills and fields, forests, open plains and villages, killing babies, children, young men, old men, fathers, sons, and mothers . . . they overthrow, they despoil, they destroy, they burn, they ravage – sinister cohort, fatal phalanx, cruel host!

The number of ships grows larger and larger, the great host of Northmen continually increases, on every hand Christians are the victims of massacres, looting, incendiarism, clear proof of which will remain as long as the world itself endures; they capture every city they pass through, and none can withstand them.

From the wrath of the Northmen, Lord, deliver us.

But these writers were all monks, who watched helplessly as the pagan raiders, whose culture and values were completely alien to them, literally rode rough-shod over Christendom. Ecclesiastics in the eighth and ninth centuries were not unfamiliar with war; what appalled them about the Vikings was their utter disregard for the sanctity of churches, from which they cheerfully plundered the considerable wealth of gold and silverware.

On the other side of the coin lies the idealized view of the Vikings created by their own poets and encapsulated

A Viking chieftain and a well equipped Viking warrior.
The arms and armour are not markedly superior to those
of the Vikings' enemies.

in the Icelandic sagas. These were not written down until the twelfth century; the narratives are probably much older and were handed on by word of mouth, but undoubtedly exaggerate the literal truth in order consciously to create the myth of the invincible Viking warrior. A tribute to one Viking prince shows some of the qualities admired: 'He could run along the oars of his ship while the crew was rowing, could play with three daggers in the air at once, and could cut and strike equally well with both hands and could cast two spears at once.' Both this and the Christian view of the Vikings are exaggerated; combined with archaeological evidence it is possible to build up a picture of Viking warriors and to appreciate their considerable achievements.

There was no Viking 'nation' as such; the Northmen did not even think of themselves as Danes, Norwegians or Swedes, but simply as members of the war-band of Skane or of Vestfold or whatever their own settlement was called. Despite their common religion, language and culture, they fought among themselves as well as against the peoples of the south. It is true that the men of different areas tended to join together and specialize in raids in a particular direction: the Swedes penetrated into eastern Europe, founded the first Russian state at Kiev, raided Constantinople (where some would later serve in the Emperor's Varangian Guard) and eventually reached Baghdad; the Norwegians concentrated on raiding Ireland, the Hebrides and Orkneys, and parts of Scotland and England, later sailing as far as Iceland, Greenland, and perhaps as far as Newfoundland. The Danes wreaked their terror mainly on the lands of the Franks and the English, although they also reached the coast of Italy and sacked Pisa.

The motives behind these extraordinary far-flung expeditions are a source of continuing debate among historians; the consensus is that pressure of population on the scarce arable land in Scandinavia, combined with the practices of polygamy and inheritance by primogeniture, created a surplus of males who had to find an outlet for their energies. Contemporary accounts seem greatly to have exaggerated the numbers involved, however, leading many scholars to believe that after the initial successful forays, large numbers of settlers were brought in, especially in Ireland, the north of England and northern France. This has been questioned, since archaeology has established that the size of the boats could not have accommodated as many men as the Chroniclers would have us believe. Another factor to be taken into account is that the fundamental mores of Scandinavian culture meant that a man had to prove himself among his fellows by indulging in exploits of ever increasing audacity and danger. Either way, as trade increased, Vikings saw the riches available in the lands to the south, and so the raids began.

After the initial forays were seen to be so successful, and resistance to them negligible because of the general dislocations of early medieval society, the Vikings embarked on what amounted to wholesale piracy, only checked when effective resistance, settlement and conversion to Christianity deprived the Vikings of the necessity to continue.

It is important to understand the workings of society in their Scandinavian homelands, if one is to explore the motivation and organization of the Viking bands. Northern society was divided into three recognizable classes which were said to have been ordained by Rig, the father of mankind, and the tenth-century poem *Rigspula* is the Viking equivalent of Genesis. The poem first describes the lowest class, the thralls who were slaves and hardly deserved to be regarded as men. Next came the *bondir* or freemen, who supplied the bulk of the fighting men. They were free farmers, owning inherited land, having equal rights under their *jarl* or king. They possessed the right to bear arms and to speak out at the local *thing* or council, which combined discussion of local issues, took decisions and administered collective justice. The bondir's savage spirit of independence often obliged them to defend their rights with axe or sword. The highest class was divided into two categories: the *jarls* or earls ranged from famous warriors who had established an ascendancy over a number of small settlements to overlords who (especially in remote areas) ruled whole provinces. Lastly, the title of king was bestowed on men of wealth, land and prowess, who often claimed to be descended from the gods. They were supposed to be elected by their peers but in practice claimants were usually confined to a particular blood-line. Their power was limited and they had no administrative machinery; they had to depend on the loyalty of their *jarls* and often had to fight off rivals. The reigns of these kings – perhaps 'war-lord' is a better term – were often glorious, bloody and brief. However, in all cases the individual Viking's loyalty only really extended as far as his immediate clan, a loyalty which was fierce and which was reflected in the warrior hierarchy. The chieftain or *jarl* fought to win, and the men fought for their chieftain.

The warrior's duties to his jarl are set out in the epic poem *Beowulf*. He swore loyalty to his lord by kneeling and placing his right hand on his sword hilt; thereafter it was his duty to protect his leader in battle. To survive one's lord was, in the Viking code of honour, a recipe for lifelong infamy and reproach. In return, the lord's bodyguard or hirdmen were clothed in silken tunics and

A Viking sword: the most cherished possession of its owner. The rich decoration and care lavished on this weapon are plain to see.

cloaks of squirrel and sable, their war-harness was the finest the lord could obtain, their necks gleamed with gold torcs, and they shared the bounties of the feasting hall. Beowulf, in his old age, looked back on his service under his lord Hygelac with grim satisfaction: 'I repaid him in battle for the treasures he gave me . . . ever would I be before him in the troop, alone in the van.'

The Viking fighting man, then, fought as part of a unit where social and military allegiance was inextricably combined, and where concepts of honour and obligation might frequently take precedence over military expediency. Indeed, in the constant emphasis on individual heroism there would seem to be a devaluing of that collective discipline which is often essential for success. In fact, however, the social cohesion of the Viking armies gave them a unity which was normally very effective (especially if they were fighting far from home) and the discipline of the Viking armies was at least as good as that of their foes from the poorly-organized states of early medieval Europe.

The Viking warrior normally wore a thick, long-sleeved woollen coat reaching half-way down the thighs; it could be belted but was probably left to hang loose most of the time. There were two kinds of trousers: the *bondir* wore long, tight versions while the nobles preferred baggy ones. Nobles also wore long capes or cloaks, with points either at the sides of the body or at the front and back. They might also wear mail-coats and helmets. The mail-coat of the noble covered the whole body and was supplied with a hood; it was hot and heavy to wear, as well as being expensive. For the ordinary warrior, the byrnie – a sleeveless, waist-length mail shirt – was sufficient. Helmets came in many varieties – some were brimmed, some conical, some had visors, others nose-guards. Many were surmounted by the figure of a boar, symbol of the god Frey, or had a boar painted on the front. The winged and the horned varieties so beloved of fiction writers were worn only by great kings and *jarls* for ceremonial occasions.

The Viking's most important weapon was his sword, which was often claimed to have had a magical origin – perhaps it had been made by the dwarfs and originally given to the god Odin. It was given a name of its own such as *Buyajubítr* (Byrnie-biter), *Hvati* (Keen), or *Langhrass* (Long-and-Sharp). Viking children were trained from an early age in the handling of weapons, especially swords, and the skill with which many warriors could wield them may well have seemed magical. The sword was a long, double-edged iron weapon, sometimes pattern-welded or inlaid, with a straight cross-guard, a flat grip, another short cross-piece, and a pyramid-shaped pommel. The blade was tapered and slightly blunted at the end, for it was a slashing instrument rather than a stabbing one. The blades of many of these swords were imported from the Frankish Rhineland, while the Norse smiths made the handles. However, the northern smiths certainly had the skill to make the whole sword at home; a really fine example could take up to a month to produce. It is not surprising that in the *Edda*, one of the major sagas, the hero Sigurd's sword *Gram* is praised seven times for its fiery gleam and its ability to sever a wisp of wool floating in the river. The Vikings loved to hear gory and semi-erotic praise of their favourite weapons, and in fact slept with their swords beside them.

A close second in popularity as a weapon was the axe. There were two main types: the older one was the *skeggox* (beard axe), which was used in the eighth century; later came the broad axe with its curved cutting edge of specially hardened iron. Whirled in a great arc, the cutting power of the axe was sufficient to decapitate a horse or split its rider – armour and all – to the saddle bow. The axe was the distinctive weapon of the blood-thirsty Viking of the horror stories.

Other weapons used by the Northmen included a seven-foot spear with a socketed head and an iron-shod butt; spear blades, like swords and axes, might be inlaid with intricate patterns of silver. Vikings of the far north used bows and arrows, the latter being sharply pointed and carried in cylindrical quivers. Finally, the warrior carried in his belt a single-edged, bone-handled knife; this was also worn by Viking women on a chain across the breast. The small, round shield was made of wood, not very thick, reinforced at the centre with an iron boss;

it was hung along the gunwales of the longship during the voyage.

Some of their foes ascribed the success of the Vikings to their superior weapons – an Irish writer claimed their victories were due to 'the excellence of their polished, ample, treble, heavy, trusty corslets; and their hard, strong, valiant swords; and their well-riveted long spears'. But this alone does not explain the length of their reign of terror. More important was their ability to raid from the sea; they had a strategic mobility which their foes lacked and their descent on a coast or up a river was often completely unexpected.

The pride of the Vikings were their longships thought by many experts to have been the most seaworthy type of

The Oseberg longship, uncovered during the excavation of a burial mound in 1904. Similar to the Oslo longship discovered in 1880, this craft shows the basic construction and narrow draught of these vessels.

A window in Canterbury Cathedral showing the siege of Canterbury by the Vikings in 1011 AD. Despite Alfred's great victories over the Vikings, they still menaced England's coastal towns during the 11th century.

basic craft ever conceived. Thanks to the discovery of one of these vessels in a funerary mound near Oslo in 1880 – miraculously intact after a thousand years – we know exactly what they looked like. The Oslo longship was a medium-sized one constructed of oak planks, 76 feet long and 17 feet in the beam; most importantly, it drew only three feet of water, allowing it to move easily up shallow creeks and streams, and thus penetrate far inland. There were sixteen oars on each side, and these were 5 to 6m (between 17 and 19 feet) long. The mast was 12m (40 feet) high, supporting a large square sail. Lining the sides of the vessel were sixty-four wooden shields, alternately yellow and black, and the prow and stern were carved to resemble a dragon's head and tail.

There were no benches for the rowers; presumably they sat on their own sea-chests.

A longship might have had a crew of between thirty-five and fifty with extra room for perhaps seven or eight horses (known to have been carried in some longships) or for prisoners or extra warriors. Kings often had much larger vessels: the *Long Serpent* of King Olaf Trygvason was said to have had thirty-four pairs of oars and an overall length of 55m (180 feet); it is supposed to have carried two hundred men into battle, (although this is unlikely) and with its gilded prow and stern-post rising 4.5m (15 feet) out of the water it was the most expensive ship of its time.

Pride was justified, even if the size of the ships was sometimes exaggerated. They could be sailed, rowed, towed and beached with ease; they even moved them on rollers across country as well, as the astonished inhabitants of Paris witnessed in 886, when a Viking band, denied access to the Seine, rolled their ships 600m (2000 feet) from one side of the town to the other. They used the ships as safe bases, too, before forts were built, mooring mid-stream in rivers or sheltered estuaries out of reach of their frustrated victims. Where rivers became too shallow even for their ships, they built dug-out canoes from logs up to eight yards long, but only twenty-six inches deep which they used, as always, to surprise their chosen targets.

This use of mobility was not limited to the sea. Horses were treasured; saddles, bridles, spurs, bits and collars were often finely ornamented and horses were sometimes buried with their masters. Although the Vikings dismounted for battle, they moved as swiftly as possible, and if horses were available, they made use of them.

A further factor in Viking success was their effective use of fortification. The military camps of the Viking raiders were surrounded by a circular rampart and ditch; inside, roads paved with wood divided the camp into quadrants. A number of wattle-and-daub houses holding 50 men each would be built in the quadrants, with storage pits beneath the ends of the houses and in the centre, a fireplace, ventilated only by a hole in the roof. Primitive as these camps sound, they were the most sophisticated ones in Europe since the fall of the Roman Empire.

Excavations of the earthworks at Trelleborg in Denmark have revealed that the Vikings must have had precise instruments for surveying, based on a standard unit of measurement. The forts they built were important to Viking strategy – in fact the word Viking comes from the Swedish *vika* meaning a place for withdrawal, usually associated with a bay or fjord, and modern place names including this element ('vik' or 'wick') show evidence of Scandinavian influence or settlement. At the same time, the meaning of the word helps to illuminate the Vikings' success: they did not simply raid indiscriminately wherever there happened to be an opportunity, but established for themselves strong bases, for instance in Ireland and the north-east of England, from which they could reach their targets easily, and just as importantly, to which they could withdraw safely, especially in winter.

Ultimately, however, the Viking warrior was so feared because his whole life, his whole social being was devoted to war, and fighting was considered the highest activity possible. Death was not feared by the Vikings; in fact, bravery, self-control and a positive disdain for death were central to the Viking warrior's code of conduct, at least within the context of his own clan or warband. Life was cheap, right from the start, when weak babies were exposed to the elements to see whether they could survive (they usually did not). This philosophy led to heavy casualties in battle since it was considered criminal to leave the fray, even when the odds were against them: Vikings literally fought to the death. There seems to have been little thought for the wounded as a result; light wounds were disdained, heavy wounding almost certainly resulted in death anyway. There appears to be only one reference to care of the wounded in all the sagas, telling of the bard Thornod, felled in battle, dragging himself to a hut where an old woman was bandaging the wounded; even here, as was the way with Viking storytelling, Thornod dies heroically, still standing, before he can be helped.

In any case, death was not the end for the Viking hero or *jarl*. A chieftain would be placed in his longship, together with ample provisions for his voyage to Valhalla, sometimes even with his servants, and the pyre would be set alight by his relatives and mourners. He was thus guaranteed a seat in this world through the heroic lines of the saga, rising ever heavenwards to Odin in the music of the bardic *skals* as his former comrades rejoiced in the feasting-hall.

The *berserkrs* were a typical product of Viking society. They fought without armour but with amazing strength and ferocity, oblivious to their own wounds, and were held in superstitious awe by their comrades. The *Yorglinga Saga* claims that the *berserkr* was inspired by Odin's rage:

> Odin could bring it about that in battle his enemies were struck with blindness, deafness or terror, so that their weapons cut no better than sticks, whereas his own men refused to wear mailcoats and fought like mad dogs or wolves, biting their shield rims; they had the strength of bears or bulls. They cut down the enemy, while neither fire nor lion could make an impression on them.

It is possible that the *berserkrs* were psychopathic or epileptic; it has been suggested that their reckless bravery was caused by drugs in a species of toadstool, or by drunkenness.

Psychological preparations were important for all Viking warriors before any major engagement. First the war-band would be roused for the battle by the leader, or their spirits raised by a minstrel; then the enemy would be doomed to destruction by the throwing of a Viking spear over their heads, consigning them to Odin, god of war and gatherer of the slain. To the accompaniment of war-cries – 'Press on, press on, prince's fighters, hard and

A Viking gravestone. Two of the predominant decorative elements are those which gave the Vikings the mobility which was their best weapon: on the top is a horse, and below that a dragon ship.

hard on farming men' was the Viking slogan at the battle of Stiklestad in 1030 – the chief would advance, surrounded by his *skaldborg* (shieldfort) of freemen. Both sides would send up a hail of arrows, followed by a volley of spears, and then the Vikings would move in with sword and axe. Tactics were elementary; the wedge was the only proper formation the Vikings seem to have used. They relied on their individual fighting prowess to carry them through to victory.

Viking armies could be defeated by determined foes, but the martial spirit of these northern warriors enabled them to rampage over north-west Europe for a considerable period before the establishment of stronger states and a more settled society curbed their activities.

Viking Weapons

The Viking warriors brought terror to much of western Europe for centuries, and yet their military success did not seem to have owed much to superiority in either tactics or weapons. They used the same methods of fighting as most of the people they were raiding, and their equipment gave them no advantage. The helmet shown below, for example, is a reconstruction of the Anglo Saxon one found at Sutton Hoo, and it is certainly as good as anything a Viking chieftain would have worn. The sword shown below would have been about three feet long; and although the Vikings were proud of their weapons, any Irish, Frankish or Saxon warrior might have carried such a weapon.

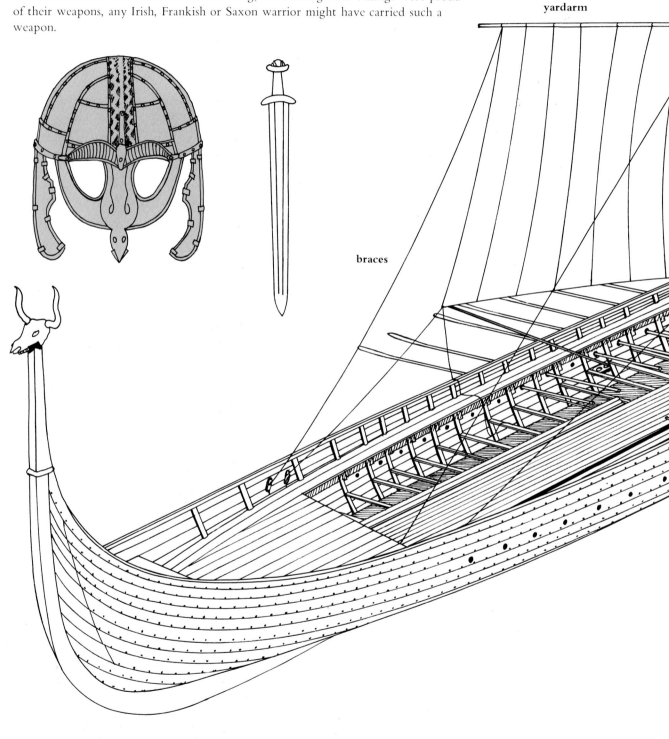

yardarm

braces

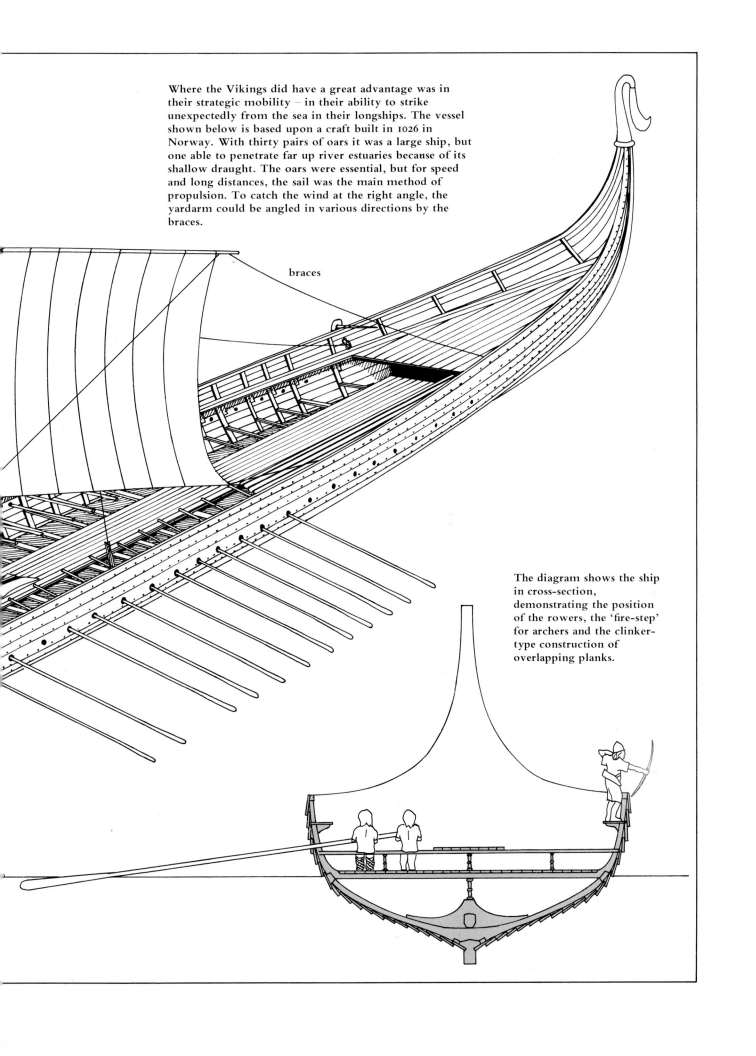

Where the Vikings did have a great advantage was in their strategic mobility – in their ability to strike unexpectedly from the sea in their longships. The vessel shown below is based upon a craft built in 1026 in Norway. With thirty pairs of oars it was a large ship, but one able to penetrate far up river estuaries because of its shallow draught. The oars were essential, but for speed and long distances, the sail was the main method of propulsion. To catch the wind at the right angle, the yardarm could be angled in various directions by the braces.

braces

The diagram shows the ship in cross-section, demonstrating the position of the rowers, the 'fire-step' for archers and the clinker-type construction of overlapping planks.

5
THE NORMANS
OF THE FIRST CRUSADE

Social organization in the early middle ages was based on war, at least in most of western Europe. The waves of barbarian migrations followed by the incessant civil wars within the emerging states meant that the old centralized institutions of state disintegrated. At the same time the assaults of militant Moslems had cut off the traditional Mediterranean sources of wealth. States no longer had the resources, troops or equipment to face the new threats from the Vikings in the north, the Magyars in the north-east or the Moors in the south, all of whom were increasingly mobile. The old infantry of free men became not only unwieldy and ineffective against sudden marine or mounted attack, but also impossible to muster, let alone to pay. The only asset the new leaders of society could offer was land. As a result, a new way of raising a new kind of fighting force developed. Starting in the Frankish empire of Charlemagne and his descendants, a system for raising an army of mounted warriors gradually came into being – the system which came to be known as feudalism.

By the eleventh century, military and economic feudalism was established in most of Christian Europe. The classic definition of the feudal knight is of a heavy mounted warrior, armed with lance and sword who, in return for a grant of land known as a fief or knight's fee, contracted to serve in the armed forces of his lord for a certain period of time (usually forty days) at his own expense. The land granted was by this time hereditary, giving the landholder the economic freedom to honour his personal obligation to his lord. Since land was the basis of wealth and political power at this time, feudal knights in fact formed the dominant social class, building castles to defend their holdings and as symbols of their power. Feudal warfare was generally a series of struggles for possession of the most important castles in a disputed territory.

Historians of the middle ages are anxious to point out the many variations to feudalism, hedging any general definition with exceptions. But in eleventh century Europe, the feudal institutions were at their most typical, and their most successful. Not only were the invasions of the Vikings and the Magyars halted; feudal forces took the offensive, and extended their influence. In 1066, the Anglo-Saxon kingdom of England was conquered by the forces of William of Normandy, Byzantine rule in south Italy was ended, and the long expansion of Germany eastwards began. But the most striking example of this aggressive feudalism was the First Crusade, proclaimed in 1095, in which an army of European knights marched through Asia Minor to capture Jerusalem and establish Christian kingdoms around it.

The fighting men who typified the martial prowess

Left A Norman Knight of the late 11th century. The armour was capable of withstanding heavy blows.
Right A church column in central France, showing soldiers with their basic arms and equipment: hauberks, nasal helmets, kite-shaped shields, spears and heavy swords.

which won these victories, and who provided the spearhead for many of the campaigns were the Normans. They embodied their age, and its contradictions, to the full. Piously Christian, yet horrendously cruel, enormously brave, but quite capable of treachery, living for war, but often unprepared to obey orders; their tangled motives and epic achievements have always fascinated historians.

The Normans were descended from Viking raiders who harried the area of the northern Carolingian empire called Neustria. When they were defeated in 911 at Chartres, their leader Rolf (or Rollo) was baptised and established by the emperor Charles III as duke of the province, to act as a buffer against further encroachments by his fellow Vikings. Rollo and his son did their job well, protecting their fief against Viking marauders – and simultaneously extending its area at the expense of neighbouring French lords. Within thirty years of the original grant to Rollo, the Normans – as his followers were called by the French – had been converted to Christianity, intermarried with the local inhabitants, and adopted the French language, but they did not lose their hunger for land and their warlike spirit. Instead they used the feudal system to build up a dynamic and hard-hitting military force which conquered England in 1066 and Sicily by 1091; at the same time, the Normans pushed the Byzantines completely out of Italy and struck across the Adriatic into Greece.

It should not be imagined, however, that the Norman knights were a unified force, fighting under a single leader. Rather, they were a social class fighting for their individual lord and their own self-interest. In the large-scale offensive expeditions of 1066 and 1091, they formed part of an army which also included mercenaries and in which they themselves had agreed to fight for periods longer than those normally prescribed for feudal levies, in the hope of land and booty.

Again, during the First Crusade, Norman knights were not a single force, but fought as the followers of various of the leaders of the Crusade. Of the nine principal leaders of the crusaders, six were Norman, leading contingents of Norman knights. Their ambitions provided the impetus which took the crusaders through the Byzantine empire, into the hostile terrain of Syria and Palestine, and enabled them to win the glittering prize they were seeking. Compared with the civilizations of the Byzantine Empire and the Seljuk Turks, the Normans were barbarians, unversed in the more subtle arts of war and diplomacy; but their vigour and the sheer fighting power of the individual knights carried them through.

The whole upbringing and training of the Norman knight was designed to prepare him for fighting. Feudal obligations ensured that the son of a knight would normally be trained to follow his father, although the training would be carried out by another relative or close friend in most cases. Riding, javelin-throwing, swordplay and charging with the lance at a shield fixed to a stake were exercises which every aspirant had to

practise repeatedly, both in order to learn fighting skills and to build up his strength – for the knight's armour and equipment could only be worn by a strong and well-built man. Anna Comnena, daughter of the Byzantine Emperor Alexius, described the armour of a Norman knight in the year 1107: 'His chief weapon of defence is a coat of mail, ring plaited into ring, and the iron fabric is so excellent that it repels arrows and keeps the wearer's flesh unhurt.' The exact composition of Norman mail has been the subject of some controversy, but it seems likely that rings were sewn onto a coat made of leather or strong linen which reached just below the knees – with a slit to allow leg movement – and had elbow-length sleeves; it weighed more than thirty pounds. This was worn over a tunic of mixed wool and linen which had sleeves reaching to the wrist; the knight's thick stockings were supported by cross-garters made of leather. Over his head and shoulders he would wear a coif of padded mail, surmounted by a simple helmet with a nose-guard but without ear-flaps, since his ears were already protected adequately by the coif. Some knights had armour made of overlapping scales rather than rings, and the wealthier ones might also have mail *chausses* to protect their legs. Men who wished to sacrifice protection for the sake of comfort and mobility might wear hauberks (mail coats) without sleeves and might discard the helmet, relying only on the mail coif as a head covering. When 'off duty' the knight would replace coif with a soft felt cap, which might be of any colour except yellow (worn

only by Jews at this time). His hair was normally kept very short, at least while on campaign.

The knight's shield was of a new type, introduced toward the end of the tenth century especially for the mounted warrior. Instead of the small round shield carried by infantrymen, this type was long enough to cover the entire left side of the knight when he was mounted; it had a semi-circular top and tapered to a point at the bottom, somewhat resembling a kite in outline. It was often made of lime-wood, covered with layers of hide, parchment and linen, and painted with fanciful designs – although systematic heraldry was not yet developed in the eleventh century. Inside the shield were a series of leather straps through which the left forearm was passed, together with a longer loop which was passed around the neck to help support the shield's awkward bulk. It could also be used to hang the shield on the knight's back if he wanted to use both hands for wielding his weapons.

The primary weapon of the knight was the lance. It was about 3m (between ten and twelve feet) long with a sharply pointed, leaf-shaped blade. In earlier times the cavalryman had thrust the lance into his enemy with his right arm, but the new technique was to couch the lance – that is, to tuck it under the arm and rely on the momentum of the horse and a firm grip instead. In order to derive maximum protection from the shield, the knight would always try to keep his enemy on the left when charging, and this meant that the lance had to be passed across the horse's neck – which also provided a bit of extra support for the lance's weight. The lance was only couched while charging, the rest of the time it was carried vertically, with the butt resting on the front of the knight's saddle.

Once the lance had been broken – often after the first charge – the knight would rely on his sword. This was longer and narrower than the Viking sword of previous centuries; a common length was about 110cm (44 inches). The guards were straight, and the pommel was shaped like a brazil nut mounted on its side. A leather thong was often laced around the grip to help prevent the knight's hand from slipping. The scabbard was attached to waistbelt, which was often worn under the hauberk, in which case there was an extra slit in the mail at hip level to accommodate the sword.

The great horse or destrier of the knight was somewhere between a modern hunter and a shire horse in terms of size and strength.

Horsemanship was a major part of a knight's training, for it was fighting on horseback which gave the knight his great advantage, enabling him to wear the heavy protective armour and carry the huge shield impossible for an infantryman. The introduction of the stirrup in the eighth century had made these developments possible. For fighting, the knight stood in his stirrups rather than sat in the saddle, and his horse had to be ready to stop instantly, to charge a solid wall of enemy cavalry, or to stand still in the heat of battle with enemies swirling all round.

Top The Bayeux tapestry has given historians a great amount of information about Norman knights, although there is considerable debate as to how reliable this is. Above A representation of Saint George killing the dragon, on Angoulême cathedral. The knight is securely balanced in his stirrups, and able to slash with his sword.

The Norman Cavalryman

The Norman knight depended for his effectiveness on his power as a mounted horseman, as an individual fighter who could deal with greater numbers of less well-armed opponents. The two elements crucial for this were his ability to wield heavy weapons on horseback, and the strength of his protective covering.

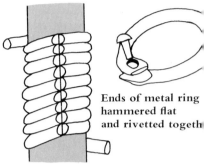

Ends of metal ring
hammered flat
and rivetted togeth[er]

Circular metal rod wrapped around
pole and then cut into individual
metal rings

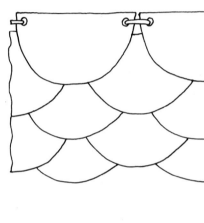

The invention of the stirrup had given horsemen a great advantage. Standing in his stirrups, and relieved of the need to grasp the horse tightly with his thighs, the knight could wield a heavy broadsword and deliver wide slashing blows to infantrymen without overbalancing. In diagram *a*, all the weight could be put safely onto the right foot to gain maximum power for the sword stroke. Alternatively (*b*) when fighting other cavalry, the knight could lean forward and sweep the sword across in front of his mount without falling forward. Finally (*c*) he could couch his lance (which would normally be rather longer than shown here) and concentrate the weight and impetus of himself and his horse onto his target.

a

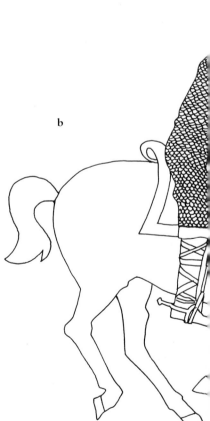

b

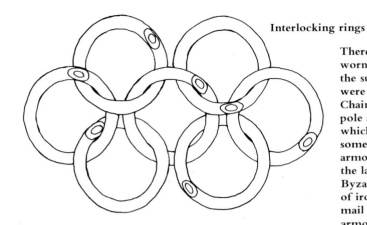

Interlocking rings

There were many varieties of mail, and the exact type worn by the Norman knight at any one time has been the subject of considerable debate. The two basic types were chain or ring mail and scalar or lamellar armour. Chain mail was constructed by winding wire around a pole and cutting through it to form a set of rings, which were then interlinked and rivetted together, or sometimes sewn onto a cloth or leather backing. Scalar armour was of various kinds. Some took the form of the lamellar protection described in the chapter on the Byzantine forces, but another method was to tie rows of iron plates together so that they overlapped. Chain mail provided a less continuous covering than scalar armour, but it was more flexible, and its protective qualities were probably just as good. Most danger to a knight came from slashing blows rather than thrusts; in this situation the gaps in the centre of the rings did not matter, and the rings themselves could be of much thicker metal than the scales of lamellar armour without increasing cost or weight unduly.

Overlapping scale armour

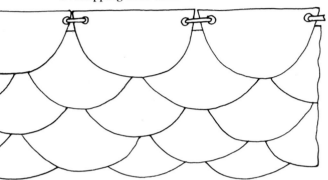

c

With lance, sword, helmet, shield and coat of mail, plus probably more than one horse for a protracted campaign, even the humblest knight could not travel without help – a team of retainers which became known as a 'lance'. There would be a shield-bearer or *escuyer* (esquire); a groom and probably a packhorse; perhaps some foot soldiers for guard duty and even one or two horsemen on less cumbersome mounts than their leader.

The heavily armed Norman knight was a formidable opponent. Anna Comnena described how 'once they have taken the initiative, they can no longer be restrained.' The Emir Onsana ibn Munquidh stated that, 'Anyone who knows anything about the Franks looks on them as beasts, outdoing all others in courage and warlike spirit, just as animals are our superiors when it comes to strength and aggression.' But this individual prowess and power was not linked to any strategic or tactical finesse: knights would merely charge *en masse* with the tactical aim of collective shock. The co-ordination necessary for the charge to be effective was often lacking and charges would degenerate into a number of individual attacks. If a concerted effort was made, it was almost never followed by a second one, because of the difficulty of regrouping. In a typical example of shrewd Byzantine analysis, Anna Comnena wrote:

> The Celts [her collective word for all the crusaders] are invincible in the first onslaught, but after that they are easily overcome because their arms and equipment are very heavy, and they behave recklessly because of their impulsive nature.

Although her analysis was quite accurate, Anna was exaggeratedly scornful. In spite of their lack of technical finesse, the Normans had such a collective pride in their ability that they could often carry on and defeat more disciplined adversaries. This collective pride came partly from their social position. Each knight had an exalted place in civil society as a warrior more than a match for foes of a lower social class. A knight's whole self-belief (and also his credibility in the eyes of society) depended on his bravery. Knightly honour could not survive the stigma of cowardice. Thus, although often outmanoeuvred, they would fight on. At the critical battle of Doryleum, Fulcher of Chartres describes how 'we [the crusaders] were all herded together like sheep in a sheepfold, trembling and frightened, eventually we were totally surrounded by the enemy'. Baldwin (one of the crusader leaders, and the son of one of William the Conqueror's tenants-in-chief) rallied the knights by inviting those who were afraid to turn back. None, of course, dared to do so publicly.

What reinforced this attitude was the fact that retreat *was* justified for social inferiors, who were not so well armed. Only the knight, the central point of society as well as the central force on the battlefield, could not retreat and had to fight to the end. Everything in the upbringing and training of the knight inculcated these attitudes; it is hardly surprising, therefore, that the knights of the Crusade, in spite of their deficiencies, were difficult to defeat in open battle. This pride in one's personal bravery and ability had several side-effects. Obviously, knights were very sensitive to anything which might be considered a slur on their honour, which contributed greatly to their difficulties in maintaining set formations on the battlefield as they broke out to find an enemy to fight if given the opportunity. They also tended to be personally vain about their appearance, a trait which the cultured Byzantines found amusing in these usually illiterate and crudely dressed northerners.

During the First Crusade, the Normans had the additional cohesion of religion. The Crusade was a Holy War, and although in this more secular age the importance of religious motives is hard to assess, piety had a large part to play in a knight's life. The Norman leaders, for example, had close family ties with the episcopacy, and enthusiastically embraced the idea of a Holy War as it suited their interests. The Normans may not seem particularly christian to us, and their values may well seem hypocritical; but there seems no reason to doubt their faith. When Robert Guiscard's forces passed near the church built by the martyr Theodore, they prayed at the church throughout the night. When the crusaders moved forward in the final advance on Jerusalem, Raymond of Toulouse led them barefoot, dressed as a pilgrim. During the siege of the city itself, the whole army walked barefoot round the walls, and fasted in response to a priest's vision.

The Holy War had its obverse side. In a cruel, warlike society, the heathen, Moslem or Jew would be given short shrift, and as the embodiments of violence, the Normans had no scruples about inflicting terror and cruelty. They beheaded prisoners and catapulted the heads into beleagured cities, and impaled heads on lances to terrorize the enemy. When they finally took Jerusalem, the inhabitants were put to the sword and the city sacked. A Moslem chronicle reported that in the area of the El-Aqsa Mosque:

> the Franks slaughtered more than 70,000 people, among them a large number of Imams and Moslem scholars, devout and ascetic men who had left their homelands to live lives of pious seclusion in the Holy Place. The Franks stripped the Dome of the Rock of more than forty silver candelabra, each of them weighing 3600 drams, and a great silver lamp weighing forty-four Syrian pounds, as well as one hundred and fifty smaller silver candelabra and more than twenty gold ones, and a great deal more booty.

In the wealth and prosperity of these civilized lands they were marching through lies another of the main motives of their advance. They wanted booty and riches unobtainable in north-east Europe. At Doryleum, they encouraged each other with a curious mixture of Christianity, social aspiration and avarice: 'Be of one mind in your belief in Christ and in the victory of the Holy Cross, because you will be rich today, if God wills,' and, 'out there on the pass, we shall either lose our heads or else become so rich in silver and gold that we shall no longer have to beg from our commanders.'

Land, however, was a more important lure. The Norman leaders were not those who had achieved success on the English or Sicilian expeditions, and there is no doubt that many of their followers too were after a fief of some kind. Indeed, Bohemund of Taranto was content when he was given control of Antioch, and he took no part in the expedition to Jerusalem. When success finally crowned the Crusade, a feudal kingdom was set up, and awards of land made.

Greed and ambition were accompanied by extreme adaptability, ruthlessness and an implacable will to succeed. The Norman leaders used whatever means they could to achieve their ends, and their followers were of the same mould. Robert Guiscard, for example, had used Saracen mercenaries during his conquest of Sicily, and refused all requests to allow conversion to be tried. The knights of the Crusade realised that they were fighting in a strange environment against a dangerous foe, and made no mistake about underestimating their enemy. One knight wrote:

Who can be so wise or learned that he is bold enough to describe the expertise, the martial virtues and the bravery of the Turks . . . then it would be impossible to find a people surpassing them in might, bravery and military genius. . . . And yet by God's grace they were beaten by our men.

They were prepared for a tough struggle, and they certainly had one. After the battle of Doryleum in 1097, for example, the crusaders continued across Anatolia through the heat of the summer:

Hunger and thirst pinched us on all sides, and there was absolutely nothing for us to eat . . . most of our cavalry ceased to exist, because many of them became foot-soldiers. For want of horses, our men used oxen in place of cavalry horses, and because of the very great need, goats, sheep and dogs served as beasts of burden.

The knights soon took to wearing linen surcoats over their mail to protect themselves from the heat – a technique they learned from the enemy. Again, during the siege of Antioch from October 1097 to May 1098, they had to endure acute privations. But here again, they hung on grimly until treachery delivered the city.

With the Norman knight of the First Crusade, then, we are dealing with a fighting man whose individual prowess was never questioned, but whose ability to function as part of an effective formation was always in doubt. Yet the elements of social pride, greed, ambition, piety, cruelty and adaptability combined to produce a formidable warrior who embodied the expansion of Christian Europe after the Dark Ages.

Experiences in the heat of the Mediterranean led to changes in equipment. By the 12th century linen surcoats worn over chainmail as here were a common feature.

6
THE FLOWERING
OF CHIVALRY

The successes of the Norman knights during the First Crusade had been due to their vigour, bravery, adaptability and complete ruthlessness. In the centuries following their success, the connection they had forged between Christianity and military success was extended in various ways. Orders of warrior monks, whose disciplined unity gave them great advantages, sprang up and became extremely powerful. The most important, and most pervasive, connection between Christianity and war, however, was in the chivalric code. By the fifteenth century, the knight had become a sanctified figure rather than a warrior, and by a steady addition of related ideas, such as those of courtly love and the Christian knight as defender of the faith, an idealized framework was built up. In practice this framework was based on pure self-interest, was selfish, hypocritical, and provided an excuse for conservatism and inefficiency.

'War is not an evil thing, but good and virtuous.' So wrote Honoré Bonet in his fifteenth-century treatise *The Tree of Battles*. 'War by its very nature seeks nothing other than to set wrong right, and turn dissension into peace in accordance with Scripture.' This idealistic view of war was commonplace among the European knightly classes during the late middle ages. The moral basis of the disputes themselves do not, however, bear close scrutiny, and there were many to scrutinize: warfare, petty or grand, was endemic in Europe during this period. Between 1337 and 1453 the English and French were embroiled in a more or less constant series of armed struggles which later historians have named the Hundred Years War. This ended just in time for the English to be able to return home for the first round of the Wars of the Roses, a thirty-year period of aristocratic conflict which began in 1455. The French also had their own civil war between the Burgundians and Armagnacs; the Italian city-states were normally at war with each other; the Spanish kingdoms fought among themselves when they were not fighting the Moors; local wars and brigandage

were endemic in the German lands. If all else failed an aspiring warrior could join a crusade in Lithuania or Spain, or even try to start his own as the Duke of Burgundy did in 1396 to drive the Turks out of Bulgaria.

Warfare in the fifteenth century was in a period of transition. The ubiquitous medieval knights were still considered, especially among their own fraternities, as central to any fighting force. Yet in practice, the successes of the archers in England and the pikemen in Switzerland, as well as the embryonic development of firearms, were all spelling the doom of the heavily armoured knight, lumbering into battle on his enormous warhorse or hacking away on foot, still fully armoured and thus hopelessly restricted both in mobility and vision. Knights as a class had come through many changes since they had led the field in the successful battles of the First Crusade.

We have already seen how in the eleventh and twelfth centuries, might really meant right, since land was the reward for military achievement and land was also the

Left *The development of armour from 14th century chainmail to the full plate of the 15th century.*
Right *The reality of late medieval warfare. A throng of men, on foot rather than mounted, clubbing and hacking at each other, with little finesse.*

source of political power. But lands gained in times of conflict were secured by the rules of heredity in times of peace; fighting became less attractive and knights preferred to pay a fee (known as a scutage in England) in lieu of personal service. Local interests gained more importance, and knights took on more administrative and judicial functions within their areas.

The thirteenth century growth of centralized government and of the merchant classes in the towns eventually challenged the knights' economic ascendancy; knighthood reacted by increasing exclusivity, exemplified in the development of heraldry (which gave visible proof of noble birth) and the élitist knightly orders and confraternities. Both of these went hand in hand with the growing cult of chivalry. At the same time, all but the most powerful nobles were becoming increasingly poorer: the expense of feudal service plus the stiff 'relief' payments that kings and overlords demanded for their recognition of the inheritance of a fief were very burdensome. It was at this point that knighthood once again ceased necessarily to be synonymous with inherited nobility.

the middle of the fourteenth century the principle of payment for military service was well-established in both England and France in all ranks of the army except the very highest. The warrior knight was in it for the money, and only differed from the mercenary soldier in that he probably had some private income to fall back on. The fief-rente, the method of payment for knights, maintained a measure of the old feudal contract. The reward for service was in direct money payments, rather like a retainer, which meant that a ruler could call on these knights at any time rather than recruiting a completely new army of mercenaries for each campaign. This had the advantage of ensuring a core of reasonably loyal and reliable veterans.

The prospect of regular pay, however, was insignificant in comparison with the potential rewards offered by the system of ransoming enemy prisoners-of-war. The higher nobility might be worth ransoms of £1000 or more to their enemies, and the price of the release of King John of France was half a million pounds – this at a time when most knights earned about £60 per year in rents from their lands. If a ransom was not paid the prisoner might spend the rest of his life in captivity, as happened to the first Duke of Bourbon; his wife could not afford the ransom, and he died twenty years later without seeing his home again. Ransoms were enforceable in the courts of law in the same way as any other debt, and could be demanded even if the prisoner was subsequently rescued by his own side. This was usually specified in the verbal agreement made when quarter was granted, and even Jean Froissart, whose *Chronicles* are full of the honourable and chivalric exploits of knights acknowledged the business of warfare, giving several examples of demands to yield, 'rescue or no rescue'. The chances of an individual soldier 'hitting the jackpot' by capturing a valuable prisoner in battle were considered to be about one in a hundred, but there was no shortage of volunteers for the army when opportunities like this were available. Soldiers even entered into agreements among themselves concerning the sharing out of profits from ransoms, rather like joint-stock companies, and ransoms could also be sold or transferred to third parties.

Regular pay and the chance of ransoms attracted both nobles and commoners to military service, but the knightly class was also imbued with the ideals of the chivalric code, a code which could only properly be put into practice in battle. The first requirement of the true knight was personal military prowess – willingness to fight, strength, endurance and heroism. The examples which the knight was to emulate were the Nine Worthies: the pagans Hector, Alexander the Great and Caesar; the Jews Joshua, David and Judas Maccabeus; and the Christians King Arthur, Charlemagne and Godfrey of Bouillon. In the court of the dukes of Burgundy where the decadence of knighthood in decline was most

Although patriotic ideals were sometimes evident, knights rarely went to war primarily to serve king and country. Nor did they go on campaign to fulfil feudal obligations of service to their lords. The feudal levy was not really a practical way of raising an expeditionary force, especially for the English, who were almost always fighting across the Channel in France; little could be done with an army composed only of noble cavalrymen who would go home after forty days. Instead, armies were raised by issuing contracts of employment and paying the fighting man the going rate for the job. By

Left *One of the subjects of medieval romance: Hercules fighting the infidel. Some of the more unpleasant aspects of medieval war – such as administering the coup de grâce by finding the weak spots in armour – are also shown.*
Right *The late medieval knight yearned for a battlefield on which he could fight how he wanted, without the problems of infantrymen, and so romanticised visions of war, as here, showed conflict between mounted men only.*

evident, there was special emphasis on the career of Hercules, whose labours were celebrated in a series of tapestries. The knight was to love fighting and adventure for its own sake, as the biographical romance of the fifteenth-century knight Jean de Bueil makes clear: 'It is a joyful thing, is war . . . when you see that your quarrel is just and your blood is fighting well, tears rise to your eye.' In theory, too, a knight could never refuse to do battle if it was offered by an enemy.

By the fifteenth century, of course, the effectiveness of the knight in battle was on the wane. They became less and less useful, but clung to their status through the feats of arms, which occasionally meant individual acts of bravery on the battlefield. (Ironically the heroic knights romanticized by the chivalric writers of the time were often commoners, soldiers of fortune like Sir Hugh Calverley or Sir Robert Knowles, who, when researched, have been found to be little more than brigands.) More often, a knight would prefer to prove his prowess in the enormously expensive displays of pageantry on the tournament field.

Another ingredient in the chivalric code was honourable treatment of fellow-knights; a defeated opponent was a social equal and should be treated as such if he had conducted himself worthily. Even a knight who turned his coat, as one from Hainault did during the Hundred Years War, was still praised by his French opponents because he fought to the last breath with his new English allies and refused to seek asylum in a nearby castle while he was being overcome. Another knightly hero, Jacques de Lalaing, was 'gentle and humble' to his enemies once they had been beaten, and gentleness to his social equals was also one of the notable attributes of Edward, the Black Prince.

Knightly courtesy, however, was only for fellow-knights; the common people were clearly excluded, since they were held to be incapable of possessing honour. It made no difference to a knight if they were starving peasants or rich burghers; all were held in contempt. René d'Anjou, in the *Livre de Tournoi*, listed categories of knights who were refused entry into tournaments; these included usurers, who, it was assumed, would have lowered themselves to the level of merchants to lend money for profit, and those deemed to have married below their station.

The clearest example of knightly contempt for the common people is the scorched-earth policy of both the English and French armies in the fourteenth and fifteenth centuries. Edward III himself gave an account of the English *chevauchée* or campaign of destruction in 1339:

On Monday, the eve of St Matthew, we left Valenciennes, and the same day the troops began burning in Cambresis and they burnt there throughout the following week, so that the country is clean and laid waste, as of corn, cattle and other goods . . . so we proceeded each day, our men together burning and destroying the country for twelve or fourteen leagues around.

A charitable mission organized by the pope found that this raid had devastated 174 parishes in four dioceses. As an anonymous (and presumably plebeian) writer remarked in 1356, 'They claim to be of noble parentage. My God! From where do they get such false wishes? What good deeds have they engaged in?' Even the practice of ransoming prisoners increased the hardships of the peasants; noble lives might be saved, but the money had to be raised by increased taxes on those who farmed the captured knight's estates.

Originally, it is true, the chivalric code had been influenced by the church, who from the time of the First Crusade had called for the protection of the poor against injustice and hardship, defence of widows and children, respect for clergy and for church property, and so on. By the end of the fourteenth century, however, the knight's duty to help the weak had been debased and romanticized into devotion of womanhood – but only women of the appropriate class were included.

The ideal exploit of the knight in poetry and romance was the rescue of a virgin, and his patron saint was

The Fifteenth-Century Knight

The chivalric code of the French knights often ran counter to efficiency on the battlefield. The concepts of honour and personal physical bravery were emphasised at the expense of discipline and tactical foresight. The battle of Agincourt in 1415 exposed the military weaknesses of the knightly code, when the smaller but better organized English army triumphed over larger French forces.

The majority of the knights and men-at-arms of the French army fought dismounted in order that their horses would not be subject to the arrows of the English longbowmen, and formed themselves into long lines, about eight ranks deep each. Between the two lines were the Genoese crossbowmen, who could therefore, play no effective role in the battle. Groups of cavalry were placed on the flanks and in the rear. The French lines soon became disorganized as the knights struggled to get into the place of honour in the front line, and the slow moving lines presented a good target to the English archers. But the real disaster was that the cramped, unwiedly mob which the French army soon became, was unable to give any scope for the warriors to use their weaposn properly, and they were butchered by the English archers and men-at-arms when the combat at close-quarters began.

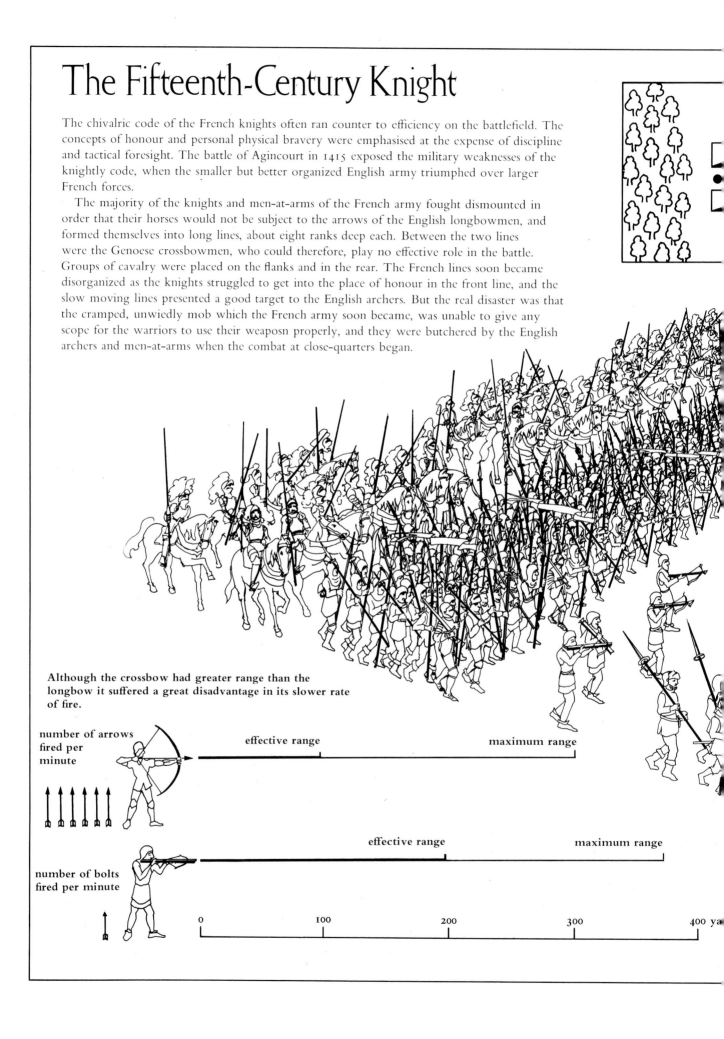

Although the crossbow had greater range than the longbow it suffered a great disadvantage in its slower rate of fire.

number of arrows fired per minute

effective range

maximum range

number of bolts fired per minute

effective range

maximum range

0   100   200   300   400 ya

cavalry

dismounted knights and men-at-arms

dismounted knights and men-at-arms

crossbowmen

helmet

bevor

pauldron

breastplate

couter

taces

vambrace

gauntlet

cuisse

poleyn

chausse or greave

soleret

By the middle of the 15th century the knight was virtually encased in metal plate armour. He had a high degree of movement, however, because of the skilful jointing of the expensive suit of armour. It would require a master craftsman to construct such armour.

George, who killed the dragon and saved the princess. Every aspirant to knighthood had to have a lady to whom he had dedicated himself, and she would rarely be his wife. This love was inevitably adulterous, whether consummated or not. It was to win her favour as well as to vindicate his own courage and honour that a knight undertook his exploits; Jean le Menigre 'served all, honoured all, for the love of one'.

In order to emphasize his military prowess or his devotion to his lady, the knight often expressed a chivalric vow. The vow involved the knight's word of honour, which had a particular mystique. These vows often went to ridiculous extremes. The famous Constable of France, Bertrand de Guesclin, once vowed that he would eat no bread until he next fought the English; the Earl of Salisbury was one of many knights who from time to time vowed to keep one eye covered until they had performed some worthy deed. In 1454, Philippe Pot vowed in the presence of his lady that he would keep his right arm uncovered and would not sit at table on Tuesdays until he had fought against the Turks.

Although the knight's lady could not accompany him to the field of battle in order to view his worthy acts of

Above *Single combat in the lists before one's monarch was a major honour, and to 'break a lance' was considered a credit to the contestants.*
Right *The richness which was often associated with the equipment of the medieval knight is shown in this Spanish painting, depicting a scene from the Crusades.*

devotion to her, the tournament gave him ample opportunity to show his prowess. Ladies were, indeed, required by the fashion of the times to attend tournaments. These had begun life as mêlées of armed knights, which were mock battles of great ferocity and represented the medieval warriors' only attempt at concerted training in arms. By the fifteenth century they had degenerated into pure spectacle, more sport than military training, and, with the increasing expense, open to an ever-narrowing stratum of society. René d'Anjou stated in 1434 that tournaments were a privilege open only to those of the rank of knight banneret or above – in other words, those who could afford the equipment and retinue required for taking part.

Knights at the tournament were arrayed in elaborate armour, and bore their impractical heraldic crests atop

Fifteenth century knights were more careful than their predecessors of their personal safety in tournaments as well as on the battlefield. By this time jousts were usually held 'at the tilt', with a wooden barrier dividing the combatants. Although horses were protected by straw mattresses and armour, the tilt avoided dangerous and expensive collisions. But the necessity of running a parallel course, if safer, also meant the knights often missed one another.

The fifteenth century tournament was a supremely theatrical event and the skills required for jousting, as well as the arms and armour worn, were completely impractical for serious warfare. Warfare in the works of Froissart and others seems little more than a series of particularly hard-fought tournaments. Needless to say the truth was harsher and more vicious. The chivalric code was deliberately fostered by both sides during the Hundred Years War to encourage reluctant knights onto the battlefield: the Order of the Garter for instance appealed directly to the knights' élitist tendencies since only a certain number were awarded at any time. This kind of national chivalry was especially necessary in England since most hereditary knights would have preferred not to leave their estates.

The conditions of battle changed quite rapidly in the fourteenth and fifteenth centuries with the increasing deployment of various types of infantry, and the armour knights wore changed too. Before about 1350, the knight wore a hauberk of mail and a jack or tunic of quilted leather or cloth; metal plates protected the upper sides of his arms and legs, and his head was concealed by a great cylindrical helm or a closer-fitting bascinet with a movable visor. In response to the havoc wreaked by infantrymen armed with longbows, however, the use of plate armour increased, with steel back and breast plates being worn over the hauberk and held in place by a tight-fitting surcoat. The shield was thus made superfluous, and the knight's armorial bearings were transferred to the surcoat. By the early fifteenth century, the back and breast plates were being made large enough to cover the entire trunk and could be joined together by straps running over the shoulders; thereafter the surcoat fell into disuse and heraldic arms were no longer displayed on the person. Metal plates now covered the arms and legs completely, not just on the upper surfaces, and a gorget was worn to protect the neck. The weak point was the armpit, which was normally protected by a roundel strapped onto the arm to cover the gap left by the curve of the breastplate. Eventually, the shoulder armour was greatly enlarged to form the pouldron and attached outside the back and breast plates; with this development, the knight was finally encased in plate armour from head to foot. At the same time, the weight of the armour had doubled since the days of the simple hauberk, jack and helm, and now stood at about 30 kilos. Unlike King Henry VIII in his jousting armour, the knight could normally mount and dismount his horse without assistance, but his movements on foot were slow and awkward. Naturally, all this armour took

their helmets. They might wear some article of green to express their hope in love or black to show their despair, or they would wear flowers as tokens of love. Here they could look forward to an approving glance from their loves as they prepared to couch their lances for an encounter in the lists. Here the valiant lover might succeed in unseating his opponent, winning his lady's favour and demonstrating his knightly bearing before a large assembly of nobles and fellow-knights.

Tournaments could only be staged by the richest nobles, since they usually took place over several days, during which time no expense was spared and each host tried to outdo the previous one in splendour. Tournaments began on Mondays and finished on Thursdays, being subject to the *treuga dei* (Truce of God) which the Church had managed to impose on the laity in the eleventh century to prevent warfare on Friday, Saturday and Sunday. However, these days were used for preparations; those taking part were vetted by the heralds for eligibility, and the assembled ladies chose a 'knight of honour' to open the proceedings.

Teams would assemble in the lists, an enclosure about 50 by 63m (160 by 200 feet), surrounded by spectator stands. After the initial charge with blunted lances, which usually broke on impact, swords (also blunted) would be drawn and the bulk of the contest was fought with them. Squires were allowed to wear defensive armour. The contest continued until dusk, after which there would be a banquet and dancing.

a long time to put on before the battle, but it was quite contrary to the rules of war to take advantage of this and charge an unprepared enemy.

The primary weapon of the knight in the earlier part of this period was the lance, although it was by no means the one which was used most in battle. The lance was made of ash, normally between ten and twelve feet in length, with a metal head; it was usually broken in the first charge and then thrown away. As echoed in the tournament, the knight then relied on his sword, most commonly the hand-and-a-half or bastard sword which was light enough to be used in one hand but could be gripped with both for a downward blow. The blade might be anywhere between a yard and 50 inches in length, and could be used for either cutting or thrusting. By about 1430, when it was usual for most of the knights to fight on foot, with flank covering of cavalry, other heavy weapons were adopted to replace the lance; these included the mace, the short-handled glaive, the battle-axe and the halberd. The knight also carried a long dagger or misericord for administering the *coup de grâce* to a disarmed enemy through the visor of the helmet or some other gap in the armour.

The knight was accompanied on campaign by servants, grooms, and perhaps even his own chaplain, as well as an esquire who aspired to knighthood himself, and normally at least six horses. On arrival in the army he would be attached to a group of between twenty-five and eighty fighting men under the command of a knight banneret – that is, a wealthy and respected knight with previous military experience who was allowed to cut the tails off his personal pennon to convert it into a square banner. Originally, the men under his command would have been his vassals, but with the substitution of paid troops for feudal levies they might come from anywhere. The constable and marshals of the army were responsible for the assignment of these companies to one of the three 'battles' – the van, centre or rear – which were both administrative and fighting units. For tactical purposes, they were often re-ordered before a fight.

The organization of knights into companies and battles was not particularly efficient, especially in the French armies. Personal devotion to the chivalric code was high among French knights and was reinforced by a great literary tradition of heroic poetry and romance, all of which made the French conservative in military matters. The same basic mistakes were made by the French at Crécy in 1346, at Poitiers ten years later, and at Agincourt in 1415; they were reluctant to fight alongside archers of their own or to co-ordinate their movements with them (or with each other). Tactics and weapons might change, but the knight preferred not to, and, in consequence, went down against the infantry of England and Switzerland. The chivalric code, growing out of feudalism, bore no relation to the needs of late medieval society; but the knights clung on to their ideals and privileges while treating war as a profitable, selfish activity, or using it as a showcase for vainglorious acts of bravery.

Opposite page *An interesting view of social relations. While the knights meet in single combat before the monarch and the richly attired court ladies, the villainous common soldiers look on.*
Above top *The perennial justification for the privileges of the knightly class was its role as a Christian influence in the struggle to recover the Holy Land, as shown here.*
Above *The siege of Calais by a 15th century artist. Infantry are not shown directly fighting knights.*

7
THE CONDOTTIERI

The word *condottiere* means 'contractor', and refers to a leader of one of the many bands of mercenaries fighting in Italy during the fourteenth and fifteenth centuries. The condottiere served his employer according to the terms of a business agreement, similar to that of any other contractor: there was a definite term of service, money in advance with further payments by instalments, an agreement not to work for the employer's competitors for a certain period after the expiration of the contract, an insurance clause covering loss of limbs in the course of duty, and sometimes even pension arrangements. The condottiere owed no allegiance to his employer except as stipulated in the contract, and this was for a fixed period only; he could just as well have worked for his employer's enemy if a better offer had been made before the signing of the agreement. He did not fight to fulfil feudal obligations, or for the honour of a dynasty or nation, or in defence of a creed; he was a true mercenary,

Left The heavy armour of the leading condottieri usually saved them from serious wounds; the main danger came from the missiles fired by crossbowmen.
Below The mercenary forces in Italy were the result of political fragmentation and a flourishing money economy.

a man whose profession was war and who was prepared to sell his services to the highest bidder.

Mercenaries have been common elements in most armies throughout military history, going back as far as the Numidian troops employed by Ramses II during his campaign against the Hittites in 1294 BC. We know that Philip II of Macedon and Alexander the Great used Greek mercenary infantry and Thessalian horsemen; Caesar made regular use of Germanic and Gallic cavalry, a practice continued by the empires in both the east and west during the following centuries, and the eleventh-century Byzantines also had their Varangian Guard of Norsemen. The force that invaded England in 1066 was principally made up of mercenaries rather than feudal levies, and Genoese crossbowmen appeared in the French armies at Crécy and Poitiers. The condottieri were preceded by the 'Free Companies' of discharged soldiers in France and followed by the famous mercenary pikemen and halberdiers of Switzerland, who provided guards for the king of France until 1792 and to the pope even today. The eighteenth century was the great age of the mercenary soldier – in 1751 only a third of Frederick the Great's men were Prussian subjects – but the nineteenth century also saw Englishmen and Frenchmen

fighting for the Spanish government. In recent years, when many had thought that mercenaries were relics of the dim past, the recruitment of volunteers from English-speaking countries for service in the Congo and Angola has led to public controversy and governmental headaches.

Despite these recent instances, the use of mercenaries has been limited to relatively small numbers, often serving as 'advisers' to newly formed armies, in most of the conflicts since the beginning of the nineteenth century. It was the French Revolution which introduced the idea of 'the nation in arms', with every able-bodied male obliged to fight for his country; previously, nations had only been too happy to hire needy foreigners to perform the dirty work of war, while apart perhaps from an élite officer class, the bulk of the citizens remained at home undisturbed and got on with their business.

A mercenary could be defined simply as a professional soldier who fights for pay. In the late middle ages, this definition certainly distinguishes him from the feudal soldier, who fought to fulfill his obligation to his lord, but it also embraces the English soldiers who were paid out of the royal exchequer for service abroad when the feudal host was inadequate – and these men certainly thought of themselves as soldiers of the English royal army. A better definition for mercenaries of the fourteenth and fifteenth century might be 'a soldier who fights for a foreign government purely or mainly for pay, for whom nationality, alliances, ideology, culture and military traditions are unimportant, and who would be equally willing to fight on the other side if hired to do so'. This was generally true in the case of the condottieri who fought for the small republics and princedoms which shared the Italian peninsula at that time.

The condottieri appeared in Italy, and only there, because Italy was unique in Europe in the fourteenth century. It was at once the richest country in Europe and the least unified; it was composed of a number of wealthy and powerful city states often at war with one another. As states they were small in area although the cities were proportionately large: Florence had the same population as Paris, while Venice and Naples had half as many again. Milan was larger – four times the size of Paris and eight times the size of London. And what they lacked in area, they made up for in wealth and influence: Milan was as wealthy as the county of Flanders; the kingdom of Naples was as rich as the kingdom of England. These cities dominated their surrounding countrysides to an extent which was unknown elsewhere in Europe, and during the fourteenth century they themselves came to be dominated by military dictators instead of oligarchies or burghers. One of the first priorities of these rulers after seizing control was the disarming of the cities' free militias organized by the city fathers and recruited from among the citizens themselves. These militias were replaced by hired forces owing allegiance to the *signore* alone; in most cases, the citizens were content with this arrangement since it meant that they would no longer have to fight. It also meant, of course, that power was in the hands of tyrants rather than in the communes themselves, but this was felt to be a fair price to pay for release from the obligation of personal military service.

In the middle years of the fourteenth century the wealth of the Italian cities and the needs of their *signori* for troops to replace the distrusted militias drew many of the French and German Free Companies across the Alps. These foreigners, the predecessors of the native Italian condottieri of later years, were soldiers who had been left without livelihoods on the conclusion of peace in their homelands; war was all they knew, and if they could not have war at home then they had to seek it elsewhere. Although each of these bands of mercenaries had their own leader, they deliberately called themselves 'companies' – that is, groups of equal companions – in contrast to the strictly hierarchical armies of feudalism. The companies comprised varying numbers of lances, each lance consisting of a knight, a squire, a page, and two archers. In the most famous group, the White Company, many of the lances included two pages, since they had extra work to do: they were required to keep the knights' armour polished 'so that when they appeared in battle array, their arms and armour gleamed like mirrors and so were all the more frightening'. This uncommon practice gave the company its name; most knights in other companies preferred to paint their armour black, since this was a much less time-consuming method of preventing rust. The White Company was led at first by a German, but after his beheading by his erstwhile employers at Perugia he was replaced by a captain known to the Italians as Giovanni Acuto – otherwise Sir John Hawkwood of Sible Redingham, Essex. Hawkwood, one of the few mercenary captains to die in his bed, is well known for his succinct summary of the mercenary's attitude: when a pair of friars saluted him with the traditional Franciscan greeting 'God give you peace', he retorted, 'Do you not know that I live by war, and that peace would ruin me?'

The White Company was only one of many such bands – there was also a Black Company, a Company of the Flowers, a Company of the Star, and several different Companies of St George. All were employed at various times by the Italian powers, but during periods of truce or peace they stayed together and they committed many atrocities; at one point the pope and the emperor worked together to try to get them sent east to fight the Turks instead of ravaging Italy. In the following year, however, during a rebellion against papal authority, the pope himself hired the White Company, which changed its name to the Holy Company and proceeded to massacre every man, woman and child after capturing the town of Cesena.

Up until now it had been the Free Companies of foreigners who had done most of the fighting, but in 1379, when the Great Schism in the papacy brought

about a new war between the rival French and Italian claimants to the chair of St Peter, the first native Italian company came into being. Alberigo da Barbiano, its leader, had learnt his trade under Hawkwood in the White Company, and he now formed his own Italian Society of Saint George to fight on behalf of Pope Urban VI against the Breton mercenaries of Urban's French rival, Clement VII. This company, to the surprise of most contemporary observers, smashed the experienced Bretons at the battle of Marino near Rome, ending foreign dominance among the mercenary troops in Italy and ushering in the new age of the Italian condottieri; Alberigo da Barbiano was soon schooling other potential captains in the art of war, and before long it was the foreigners who learned their trade at Italian hands rather than *vice versa*.

The concept of a society of equal companions was not so important among the Italians, who had fewer feudal ties to react against than the northerners; their mercenary bands were more rigidly hierarchical, and the word 'company' passed out of use around 1400. The forces of the condottieri were organized in such a way that the leader's own unit, the *casa*, or household, was always the largest, reflecting his importance in the command structure, and it also included trumpeters, chancellors and grooms. The casa would contain the leader's most trustworthy men-at-arms, and there was a semi-feudal element in their attachment to their captain general. Many of these leaders were independent rulers in their own right and their men-at-arms would include members of noble families from the soldier prince's domain. This had several advantages: it gave an air of respectability to the troop, there was more likely to be personal loyalty to the leader and so less likelihood of treachery, and it would also guarantee that the aristocratic families from which these soldiers were drawn were less likely to stir up trouble at home while their prince was away on campaign.

A late 14th century battle in Italy. The various bands have different banners, and drummers and trumpeters are sounding the charge.

ment. He would collect reports from diplomats and spies as to the quality of various companies, and the availability of the best captains. By the mid-fifteenth century trustworthiness, good organization, stability and loyalty, together with a creditable list of victories, were considered more important than outstanding bravery, skill and brilliant success, since these might also bring over-ambitious or reckless captains who would cause too many problems to the contracting state. Individual soldiers were also recruited, contracted direct to the state rather than through a particular captain.

Recruitment could be a tricky business, especially with the constantly shifting political alliances in Italy at the time. There were fierce personal rivalries between the condottieri and some men would not serve in the same army, let alone agree to take orders from one another. In the southern states, where civil war was constant and condottieri often acted as arbiters, they could actually affect policy: Naples was forced into attacking the army of Sigismondo Malatesta by the city's captain general, Federigo da Montefeltro, who happened to be Sigismondo's arch enemy. The reverse was also true, and the collaterale had to be careful not to set armies recruited from the same area against one another, because of possible close family ties which could lead to mutiny.

Once recruitment was complete, muster rolls of every man, horse and piece of equipment were written up. The efficiency with which these were compiled directly affected the discipline of the army since proper inspections and correct payment were impossible without them. The system seems to have worked with varying success. States with permanent officials were generally more efficient than those, like Florence, still content with temporary collaterali. A Milanese commander complained of the state of his Florentine volunteers in the 1430s:

> These Florentine troops are so badly organized that it disgusts me; the men-at-arms are spread out in confusion, often with the squadrons mixed up together in a way which seems to conform to no plan, and with squadrons as much as a mile apart. The soldiers are billeted all over the place without any provision for pioneers or other essential auxiliaries; there are very few infantry, about 700, of which 150 only are properly armed, although I have made constant protests about this.

In theory, the conditions of service laid down for the condottieri were not unreasonable, although in practice it rarely worked out quite so well. Fair prices for food were often specified in the soldier's contracts, but as one commander complained:

> The Florentine officials sell victuals at the dearest price possible, without any concern for the regulations con-

Like the rest of the band, the casa was made up of squadrons of five lances commanded by corporals. Five of these squadrons made up a troop, which was led by a constable, and this was the primary tactical unit in time of battle. In 1441, the band of the condottiere Micheletto Attendolo, a typical leader, comprised 561 lances in all.

By the fifteenth century the administration of the army had become an important element, and permanent officials, *collaterali*, were appointed to draw up the contracts or *condotta*, supervise inspections of the troops to make sure there was a full muster, recruit, pay and demobilize troops as necessary, apprehend deserters and enforce disciplinary regulations. The collaterale could be as powerful as the captain general of an army. Collaterali began as temporary officials, appointed as and when needed; but they eventually became permanent, especially when the condottieri were powerful rulers in their own right, like the Viscontis and Sforzas in Milan.

The primary function of the collaterale was recruit-

The Condottieri

The period of the condottieri in Italy saw a continued development in armour; the workshops of northern Italy and southern Germany turned out some of the greatest examples of the armourer's art. Three styles are shown below. The period also saw the gradual development of firearms, which in the end were to change the nature of warfare utterly, and leave the condottieri outmoded. The crossbow remained the most important missile weapon, however, and in the static siege warfare which many campaigns consisted of, it was the most deadly piece of equipment in use.

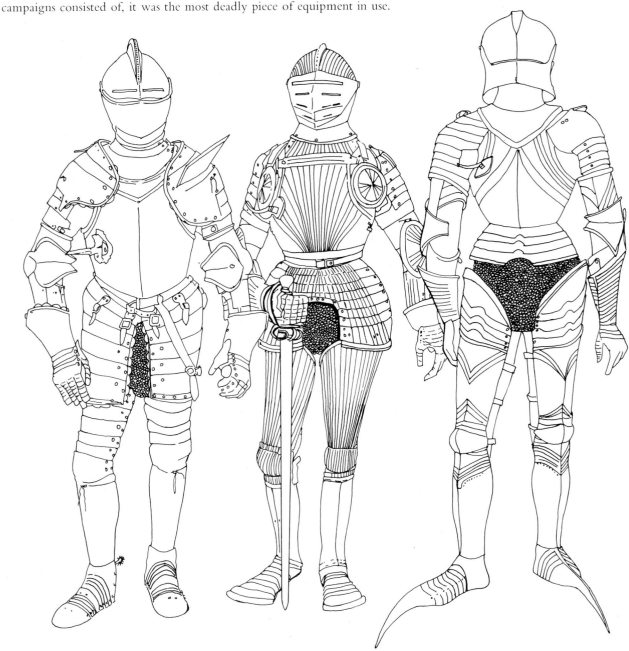

Left: Italian armour dating from 1500, made by craftsmen of Brescia for the Viscontis, a powerful condottieri family. Centre: Maximilian armour of the late 15th century, characterized by its fluted shapes which increased overall strength and helped to deflect blows. The helm has a 'bellows' visor. Right: Late 15th century Gothic armour from Germany, a lighter and more graceful style than the Maximilian. The popular sallet helmet is worn and the sabatons on the feet reflect the civilian fashions of the day.

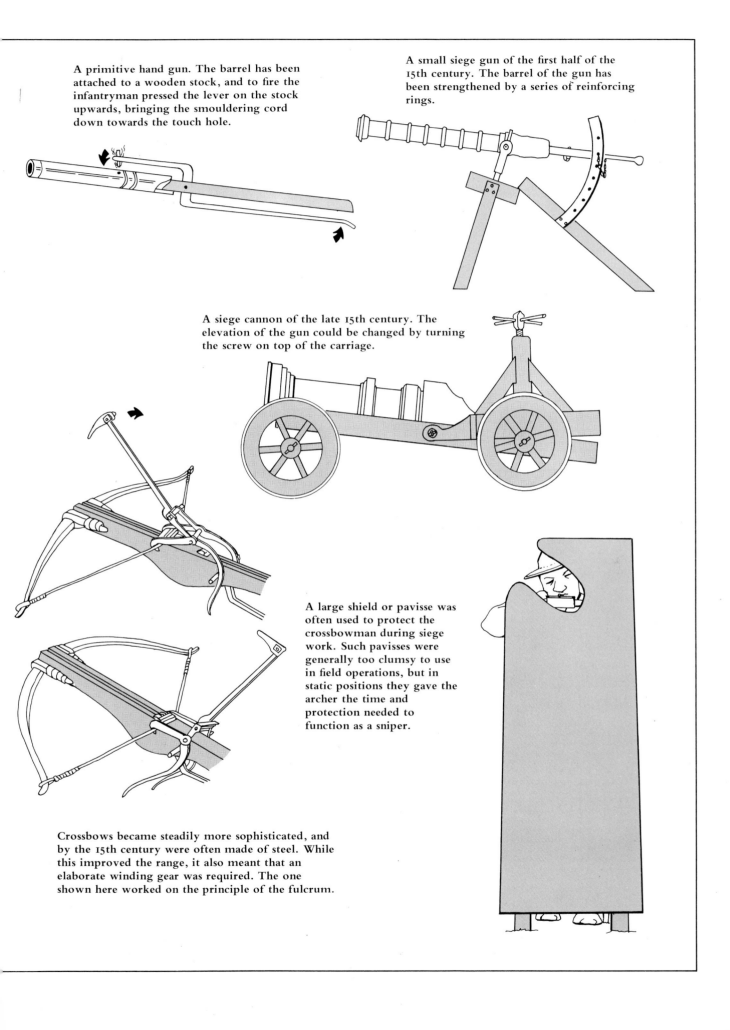

A primitive hand gun. The barrel has been attached to a wooden stock, and to fire the infantryman pressed the lever on the stock upwards, bringing the smouldering cord down towards the touch hole.

A small siege gun of the first half of the 15th century. The barrel of the gun has been strengthened by a series of reinforcing rings.

A siege cannon of the late 15th century. The elevation of the gun could be changed by turning the screw on top of the carriage.

A large shield or pavisse was often used to protect the crossbowman during siege work. Such pavisses were generally too clumsy to use in field operations, but in static positions they gave the archer the time and protection needed to function as a sniper.

Crossbows became steadily more sophisticated, and by the 15th century were often made of steel. While this improved the range, it also meant that an elaborate winding gear was required. The one shown here worked on the principle of the fulcrum.

cerning prices and quality; the money is so debased that it buys very little; and if provisions are sent from Lombardy or elsewhere these Florentines make us pay duty on them, or keep them for themselves.

Compared to other armies of the time, however, provisioning was quite well-organized. The Frenchman Philippe de Commynes wrote, 'As for the provision of food supplies and other things necessary for maintaining an army in the field, they do it much better than we do'.

This only applied while the men were on friendly territory; away on campaign they were expected to live off the land, eating whatever they could obtain by theft or violence. Similarly, the organization of billets was generally left to the men themselves. They were not billeted in the cities, although occasionally they might be billeted in small towns or villages. On the whole, they were expected to provide their own shelter: officers might have elaborate tents of various sizes according to their status, but the ordinary fighting man had to construct his own simple hut or merely sleep under his horse with his cloak wrapped around him. Winter quarters were organized with more care by a specified

Below Bartolomeo Colleoni, one of the most successful condottieri, examining presents he will give to the King of Denmark who is about to visit him.
Right A fanciful battle scene, but one which shows the weapons (if not the armour) of condottieri warfare.

official, but the condottieri tried to avoid keeping large numbers of men quartered in cities even then.

The second important function of the collaterale was to prevent desertion or insubordination, which were usually due to irregularity of payment or some other breach of contract on the part of the employing state. Desertion to the other side after contracts were signed was not normally a problem. There were two methods of paying the troops, of which the less satisfactory was for the employer simply to give a lump sum to the captain. This was in accordance with the principle that the unit was the private property of its leader, who was responsible for its upkeep and good performance in order to assure a good return on his investment.

Naturally, as much of the lump sum payment as possible went straight into the captain's pocket, and the men often suffered as a result. The alternative method of payment, whereby each man was paid a fixed wage each month, was more attractive to volunteers and therefore became more common. The wages were specified in the original contract, with occasional bonuses; in general, it appears that the commune of Florence gave the best rates and Milan and Venice the worst. Throughout the period, however, rates of pay continued to decline as the supply of mercenaries increased. In 1371, mercenaries fighting for the pope had been paid eighteen florins per lance per month; by 1404 this had been reduced to fifteen; in 1414 it was twelve or thirteen and by 1430 it had fallen as low

as nine florins since there were now so many more applicants for the job. Occasionally pay was in kind, both food and clothing, which could solve the awkward problem of provisioning.

In addition to regular pay and occasional bonuses, the individual soldier could count on special rewards for outstanding acts of bravery. In the war over Ferrara in the 1480s, for example – a war which saw the pope and the Venetians ranged against Milan, Florence and Naples – Venice promised to pay 300 ducats, plus a pension for life, to the first man who should enter the town of Rovigo. On the other hand, extra money could sometimes be squeezed out of the employers by the threat of desertion to the enemy, or at least by refusal to leave

IVA·VIRTVTE·MOTI···EDITIOREM
HERONVM·TE·PLAVDENES·EXCIPIMVS

winter quarters and resume the campaign. This method was also used by disgruntled troops to extract money owed in arrears. Most states were notoriously bad at paying on time, especially as the citizens of some of the city-states had a deep distrust of standing armies. Finally, the payment of pensions became more common as the fifteenth century progressed and mercenaries began serving under long-term contracts. Ferrando da Spagna, an infantry corporal who had served eighteen years in the Venetian service, had his right arm shot off in battle in 1446; he was offered a pension of six lire per month for life, but preferred to continue fighting with his left arm and was awarded a lump sum of 40 ducats in lieu.

Whether the employer paid the troops individually or by a fixed sum to the captain, it was in his interest to inspect the troops from time to time, making certain that the numbers and equipment of the unit were in accordance with the contract. The captains, however, disliked such inspections intensely, especially while on campaign, as it was much easier and cheaper to find replacements, both in men and equipment, when the unit was in settled quarters. Condottieri would be fined if their units failed to pass inspection; in 1468, among the troops employed by the pope, Giovanni Conti's company was thirteen horses short of its contracted strength of 200 and Niccolo da Bologna only had thirty-eight infantry instead of the required fifty.

The importance of infantry increased during the fourteenth century, even though Italy had, in Milan, the

Francesco Sforza and other condottieri. Sforza climaxed his military career by taking over Milan, and founding the Sforza line of dukes.

most important centre for production of sophisticated armour for men and horses. The quantities of armour produced early in the century were enormous: in 1427, after a defeat by the Venetians, the Milanese army was stripped of all its armour, only to be completely re-equipped with enough for 4000 cavalry and 2000 infantry in a matter of a few days. The type of armour worn in this period was carefully fitted to allow as much freedom of movement as possible, and weighed between 20 and 25 kilograms plus another two kilograms for the helmet – comparatively light and strong but stifling during a long day under the Italian sun. In addition, it was becoming more common for infantry to attack the horses, forcing their riders to dismount and fight on foot. But the most important factor in the decline of the armoured cavalry was the great increase in the use of hand-guns, whose bullets could penetrate even the best armour of the period unless its weight was greatly increased, thereby sacrificing mobility once again.

Infantry units armed with guns began to be formed increasingly among the condottieri themselves in the latter half of the fifteenth century; there were 1250 hand guns and 352 arquebuses issued to the Milanese in 1482, and only 233 crossbows. But until then the crossbow had been the most important missile weapon and companies of English archers were employed specially; as successors to the famed John Hawkwood they commanded better rates of pay than their Italian counterparts.

The use of artillery was also associated with the condottieri, in association with the development of techniques of field fortification in the fifteenth century. But generally only the most sophisticated companies had their own artillery, as the resources for manufacture and organization of large guns was only possible for the employer states.

Each condottiere captain taught his followers his own style of warfare, but they were all ultimately dependent on one of the two schools of fighting established by the two most important pupils of Alberigo da Barbiano in the late fourteenth century. The *bracceschi*, followers of Braccio Fortebraccio, relied on the momentum of assaults like hammer-blows; it is not surprising that Braccio himself was described as 'a fine figure of a man, except that he was scarred all over'. The *sforzeschi*, on the other hand, followers of Muzio Attendolo Sforza, were renowned for their skill at manoeuvre and their good use of battlefield tactics. These schools of fighting took over where the old foreign companies left off, and even before the end of the fourteenth century it was claimed that Italy was 'the best possible school of warfare' and that 'he who has not fought in Italy is of no military use anywhere'.

This eulogistic attitude was not universal, however, and was not common in non-military circles. The member of a citizen militia, a national army, or a feudal host might be willing to die for his homeland or his lord or his honour, but a soldier who fought only to make a living was obviously going to be reluctant to die at it. The condottieri certainly avoided sieges, the most dangerous type of warfare at that time, preferring to fight in the open field where the soldiers' armour offered them quite good chances of survival. Nicolo Machiavelli, the author of *The Prince*, summed up:

> They directed all their efforts to ridding themselves and their soldiers of any cause for fear or need for exertion; instead of fighting to the death in their scrimmages they took prisoners without demanding ransom. They never attacked garrison towns at night, and if they were besieged they never made a sortie; they did not bother to fortify their camps with stockades or ditches; they never campaigned in winter.

Another politician wrote that the condottieri 'are our natural enemies, and despoil all of us; their only thought is to keep the upper hand and to drain our wealth'. These views may be one-sided, even exaggerated, but that there was a certain amount of justice in them is illustrated by the great battle between the forces of the condottieri Niccolo Piccinino and the Count of Carmagnola, in which the latter captured 5000 cavalry and a similar number of infantry without the death of a single soldier on either side, 'though the slaughter of horses was incredible'. But it must also be remembered that the employer states wanted cheap, quick victories, whereas to the condottieri it was their livelihood, and they had

every intention of staying alive.

If death in battle was relatively uncommon, there were still plenty of other dangers. Condottieri resorted to trickery in much the same way as the Byzantine army. The poisoning of the enemy's wells or food supplies was considered acceptable behaviour, as well as being safer than battle, and other forms of 'biological warfare' included the catapulting of diseased corpses into walled cities during sieges. Betrayals, especially at the hands of their own employers, were an added risk for the condottieri leaders; the Venetians were especially known for this. Their attempts to poison the condottieri Francesco Sforza and Filippo Maria Visconti were unsuccessful, but they dealt with the Count of Carmagnola in a more forthright manner. He was arrested on the evening following his triumphal entry into the city after winning a string of victories, tortured for twenty days, and finally dragged out to the Piazza San Marco and beheaded. The charge against him was never revealed. Even Venice, however, could not keep up with the record of treachery established by the pope's son Cesare Borgia, who had three of his former employees strangled at one time after calling them together for a conference at one of his castles.

The rewards for the condottiere who succeeded in escaping death in battle, assassins, poison, spies and traitors could, however, be substantial. Under the terms of most contracts, a condottiere was allowed to keep all movable goods captured in enemy territory, although they were fined for looting property on friendly soil. Castles and land went to the employer. But the ambitions of a number of mercenary leaders went higher than this, especially during the latter part of the fifteenth century. Muzio Attendolo Sforza tried unsuccessfully to seize control of the kingdom of Naples, and his son Francesco ultimately realized his ambition to rule Milan and founded the Sforza line of dukes. Braccio Fortebraccio controlled Umbria, although his state did not survive his death. Others married into noble families; Sir John Hawkwood married a daughter of one of the Visconti dukes of Milan, while the last duke wed the widow of another mercenary leader and thereby gained a dowry of half a million florins.

The warfare practised by the condottieri could, however, only continue so long as both armies played the same game, implicitly agreeing to take their employer's money and keeping losses as low as possible. It finally came to an end in 1494 when Charles VIII of France invaded Italy and swept down the peninsula with devastating success. The condottieri's outdated cavalry tactics proved no match for the novel menace of the Swiss pike square. Contemporaries like Machiavelli blamed the condottieri for failing to defend Italy in the pursuit of their own interest, but in fact one of the major employers, the Milanese, began fighting on the French side. In fact the political divisions in Italy which brought the condottieri into existence in the first place caused their downfall in the end, as the disunited Italian states crumbled in the face of a determined foreign attacker.

8
FROM MERCENARIES
TO STANDING ARMY

During the fifteenth century, the mounted knight, whose weaknesses as a fighting soldier we have examined, and the experienced, but limited, armies of the condottieri in Italy were made obsolete by the tough, ruthless infantry of the Swiss cantons, and their emulators the German *Landsknechts*. But what is most interesting from the point of view of the development of the fighting man is how these mercenary forces rapidly lost their supremacy on the battlefield to the Spanish army, a force organized on a new, national basis, which was to be the model for European armies in the centuries to come.

The Swiss held sway as the finest soldiers in Europe because of their speed on the march and their undoubted fighting qualities. Machiavelli in his *Art of War* says of them: 'No troops were ever more expeditious on the march or in forming themselves for battle, because they are not overloaded with armour'. Their mobility allowed them to outmanoeuvre their opponents and force a battle on them when they were unprepared. The lightness of their equipment was at first due to the poverty of the Swiss cantons. By the fifteenth century a Swiss mercenary might wear a steel cap and a breastplate and perhaps arm pieces or tassets. But most preferred to wear no armour at all and would rely on their own skill with their weapons in the attack as the best form of defence. Most men marched into battle wearing only a leather jerkin or a buff coat. Only the officers wore closed helms, gorgets or leg armour; these rode to battle on horses but dismounted to fight.

The Swiss fought in deep columns with long pikes and halberds, a formation rather like the Macedonian phalanx – a bristling hedgehog of pikes. The halberd was not so long as the pike, though still 2.5 metres long, with a heavy steel head, tapering to a point; at the front of this could be found a blade like a hatchet, and at the back a strong spike or hook; the latter was used to catch and pull

down the reins of charging horses. At the Battle of Nancy, it was a halberd that brought down Charles Duke of Burgundy with one blow and split his skull from the temple to the jaw. The pikemen and halberdiers were accompanied by light infantry who carried cross bows and later 'hand guns'. The men who carried this weapon were called arbalesters, and preceded the phalanx of pikemen to draw the enemy's fire away from them. The Swiss were often mean, petty and insubordinate; but they were also completely ruthless. After the battle of Novara they slaughtered several hundred German mercenaries that they had taken prisoner – no echo here of the genteel and businesslike methods of the Italian condottieri.

Left A Swiss halberdier and a Spanish infantryman. The Swiss, with his salet helmet hanging from his belt, is comparatively well armoured, and dates from 1450. The Spaniard is wearing mid-16th century equipment.
Right Swiss pikemen versus the flower of Burgundian chivalry at the battle of Granson, 1476.

Though brave, however, the Swiss gradually lost their earlier effectiveness when used as mercenaries. Their first battles had been in defence of their homeland, and although they were always a fearsome foe, they were much less reliable when fighting for money. Their system of command, for example, was ill-suited to an extended campaign abroad. The commanders, the captains, were elected by their men; they were usually distinguished veterans from earlier campaigns, but apart from this there was little difference between officers and men. The infantry were also spoiled by success, and reposing too much confidence in their own judgement and not enough in that of their officers, often disregarded orders that they disliked. When things went badly they were the first to turn against their officers rather than give them loyal support. At the battle of Bicocca the Swiss were hard pressed and the cry went up: 'Where are the officers, the pensioners, the double-pay men? Let them come out and earn their money fairly for once: they shall fight in the front rank today.' The officers conceded to this impertinent demand, and the leader of the Swiss mercenaries, Winkelried of Unterwalden was quickly to die.

Inter-cantonal jealousies further exacerbated the tensions. Troops from one canton objected to being placed under the command of an officer from another. Neither was there any unity of command, for the officers formed a cantonal council which was usually riddled with jealousy and feuds. The end result was to leave the pikemen in the ranks on a very free rein.

This was unfortunate for Swiss military supremacy. Inspired by the success of the Swiss mercenaries, the Emperor Maximilian I formed a standing force of infantry and trained them along the same lines as the Swiss. These were called *Landsknechts*, or men of the plains, as opposed to the Swiss, the men of the mountains. They combined pikes and halberds and were divided into companies of 300 men each commanded by a captain, with lieutenants and ensigns to help him. By the end of the reign of the Emperor Charles V the Landsknecht had laid the basis of the modern regimental system. The soldiers mainly came from the Rhineland and Swabia. Their pikes were shorter than those of the Swiss, about ten feet long and were handled at the end in order to give the pike maximum thrust. They charged *en masse* like the Swiss, but unlike them, the Landsknechts held their pikes at their stomach rather than at chest height, to give a better balance of weight. Infantry carrying arquebuses advanced with them. All the infantry were equipped with a short stabbing sword called a *Katzbalger*.

Battles between the Landsknechts and the Swiss were hotly contested and brutal. No quarter was given nor expected. The Germans were determined to show that their courage was not a whit inferior to that of the Swiss. When they met in battle, the shock of the two columns crashing together was tremendous. There are tales that the whole front rank of each phalanx was sometimes cut down in the first clash; but their comrades in the second row merely stepped over their bodies and fought on. As the pikemen of the two sides clinched, the halberdiers were expected to move forward, get behind the pikemen and hack and hew with their halberds. The carnage was often frightful. At the battle of Novara, the Landsknechts lost half their strength.

In the end their superior discipline won the day. The Landsknecht system of standing armies was adopted throughout Europe, and in Spain Gonzalo de Córdoba based his tactics on it. He equipped his men, like the Roman legionary, with the short thrusting sword, a buckler, and a light shield, to fend off the pike. The Spanish infantryman wore a round steel cap, a breastplate, backplate, and greaves; at the battle of Marignano (1515) they routed the Swiss and were henceforth supreme in Europe. but although technical and tactical advantage was important in the Spanish success, the length of their sway and the extent of their success was due to superior discipline and superior organization, on a permanent basis by a nation state. The Spanish empire was the richest and most extensive in the western world during the sixteenth and early seventeenth centuries. Originally created by a series of marriage alliances, it was held together and expanded by a military machine which was unique at that time. The Spanish army was the first to be organized by the state; not quite a national army, since foreign mercenaries vastly outnumbered the Spanish contingents, but all were employed directly by the state, in an institution which was centrally regulated. It did not always work perfectly in practice since neither the communications system of the time, nor the discipline, quite matched the organizational theory behind it. But the Spaniard was undoubtedly the premier soldier of the sixteenth century.

Pierre Brantôme, viewing a Spanish expeditionary force in 1566, described it as a 'fine company of gallant and valiant soldiers on the march . . . all were seasoned veterans and so well appointed as to uniforms and arms that one took them for captains rather than private soldiers . . . one would have said princes, rather, so stiff they were, so arrogantly and gracefully they marched'. They were flexible enough to adapt to new developments in the use of artillery and hand-guns, and this allowed them to triumph over the Swiss, who had dominated the military scene in the previous century. As agents of Europe's most dynamic and aggressive nation, they were called upon to fight in North Africa, Italy, France, the Netherlands, and America, as well as in Spain itself. Of these campaigns, the greatest was the eighty-year struggle against the Dutch rebels (and sometimes the French as well) in the Low Countries – a war of sieges and guerrilla attacks where the Spanish regulars suffered

Above right *The horror of close-quarters infantry fighting in the early 15th century – a murderous struggle.*
Right *The various elements of infantry warfare successfully balanced by the Spaniards: longbows, crossbows, pikes, swords, halberds, handguns, artillery, and even grenades.*

many disadvantages. Another army would probably have given up the fight long before; it is to the Spaniards' credit that they carried on as long as they did.

The year 1492 was marked by two momentous events for Spain: the discovery of new lands and riches to be seized in the western hemisphere, and the defeat of the last remaining Moorish state in the Iberian peninsula. Now, Spaniards could turn their attention overseas, and an expeditionary force under 'El Gran Capitan', Gonzalo de Córdoba, was raised soon after for service against the French in Italy. From this time onwards, the Spaniard could genuinely look upon military service as a life-long career; Spanish armies were always fighting somewhere, usually in several places at once, for the next century and a half.

One of the reasons for the Spanish army's exceptional character in the sixteenth century context was that a large part of its recruitment was by conscription. The Ordinance of Valladolid in 1494 decreed that one in twelve men between the ages of twenty and forty-five was liable for paid service at home and abroad. The French had also tried conscription for their *gens d'armes*, but had never succeeded in putting it into practice. The Spaniards did, and the conscripts created a solid core of professionals with standards which remained high even when the army, desperate for manpower as the area it had to defend grew ever larger, began to rely more and more on volunteers from all over Europe.

Volunteers from within Spain were always welcome, however, and regular recruiting drives took place. A captain, having obtained his commission from the king, would march with his ensign, corporals and drummer into the towns and villages listed in the commission and set up his recruiting headquarters at the local inn. Men between the ages of sixteen and fifty would be invited to join up, and there was usually little difficulty in recruiting a sufficient number in areas where food was scarce. But it was not only the impoverished and unemployed who were enlisted; gentlemen of birth, too, were sought for each company. As the Duke of Alva wrote, 'Soldiers of this calibre are the men who win victory in the actions, and with whom the general establishes the requisite discipline among the troops. In our nation nothing is more important than to introduce gentlemen and men of substance into the infantry so that all is not left in the hands of labourers and lackeys.' In Spain as in other European countries, military service was an acceptable career for impoverished gentry who were forbidden by social custom to enter into trade. In addition, there was no stultifying chivalric tradition in Spain; it was no dishonour for a gentleman of noble birth to serve in the ranks of the infantry. Additionally, heavy horses had never been suitable for the dry plains of Spain, skirmishing against the Moors had required fast, light mounts. The heavily armoured cavalryman which was the ideal of the French nobility, for example, had never developed to the same degree in Spain.

Spain's manpower resources did not match her military commitments, so subjects of the Crown from outside the Iberian peninsula had to be recruited to fill up the ranks, even though such men were not as highly regarded as the Spaniards themselves. Italians from Spanish Lombardy, Naples or Sicily were almost as good as Spaniards, but Burgundians and men of Alsace were considered third-rate, and the Germans and Walloons even worse. The Spaniards enjoyed better pay and conditions of service than these foreigners, and were known as 'the sinews of the army'. A French observer wrote of the Spanish army in 1596, 'As for the Spaniards, one cannot deny that they are the best soldiers in the world, but there are so few of them that scarcely five or six thousand can be raised at a time'. At its greatest strength, in 1640, the army in the Netherlands included more than 17,000 Spaniards – but it also included more than 37,000 loyal Netherlanders, 15,000 Germans, 4000 Italians, and even 2500 Britons. By this time, however, the Spanish army was already well past its best.

Once a captain had recruited his company, it would be assigned to a *tercio*. This was a Spanish innovation of the 1530s by which twelve companies of about two hundred and fifty men each were brigaded together into a permanent administrative and tactical unit, the ancestor of the modern regiment. Tercios were formed on the basis of geographical origin – one of the earliest to be formed was *Lombardia* – but they often won individual nicknames as well, such as 'The Immortals'. The tercio was led by the *maestre de campo* (colonel). Under the *maestre* there was a proliferation of officers: each company had its own captain, and the companies were further sub-divided under subalterns and sergeants. This was expensive of course, but helped to give the tercio its manoeuvrability. Small groups were allowed to use their own initiative which, with their tradition of individual bravery, the Spaniards used to good effect.

There were also many other posts of officer: the sergeant-major and his two assistants, surgeon-major, chaplain-major, drum-major, and quartermaster-major, judge-advocate, provost-marshal – and hangman.

At the start of the sixteenth century, there were about two hundred pikemen and twenty arquebusiers, together with eleven officers (including a piper and a barber) in a normal company, but two companies in each tercio were composed of arquebusiers alone.

It was Gonzalo de Córdoba who effected the change from swords and bucklers to arquebuses during the Italian wars, in response to the Swiss pikemen. The proportion of shot increased throughout the sixteenth and seventeenth centuries, and by the 1630s there were as many musketeers and arquebusiers as pikemen.

After the muster-master had examined the captain's list of recruits and ensured that all the men were sane, unmarried and of the appropriate age, the articles of war were read out. Each man had to raise his right hand and swear to obey them, and the penalties for disobedience

Gonzalo de Córdoba, the leader who earned his nickname of 'El Gran Capitan' by his thorough reform of the Spanish infantry which presaged a century of success.

The Tercio: Weapons and Tactics

The Spanish army was the leading military power of the 16th century and much of its success in battle was due to the well trained infantry, organised into tercios. The tercio was a massive formation – its size varied between 3000 to 6000 men – and combined the shock action of the pike with the firepower of the new handguns. Each tercio had a central core of pikemen surrounded by a 'sleeve' of arquebusiers who also provided the 'squares' at the corners of the tercio (one of these 'squares' is illustrated in the diagram on the right).

Towards the end of the 16th century the arquebus was replaced by the matchlock musket which increased the rate of fire and gave the musketeer a more prominent role on the battlefield.

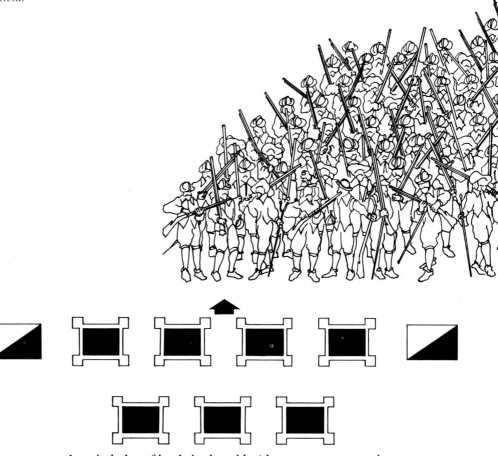

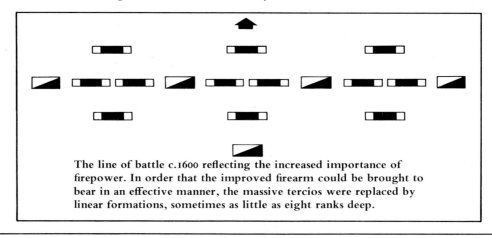

A typical plan of battle in the mid-16th century: seven tercios arranged in two lines with cavalry on the flanks.

The line of battle c.1600 reflecting the increased importance of firepower. In order that the improved firearm could be brought to bear in an effective manner, the massive tercios were replaced by linear formations, sometimes as little as eight ranks deep.

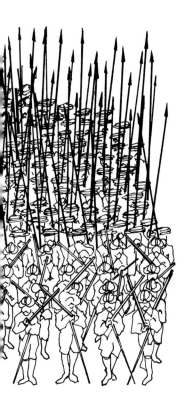

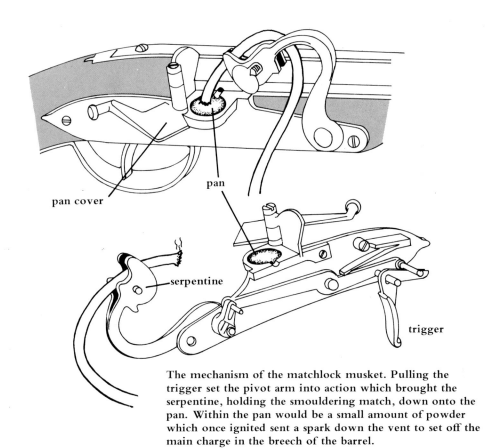

pan cover

pan

serpentine

trigger

The mechanism of the matchlock musket. Pulling the trigger set the pivot arm into action which brought the serpentine, holding the smouldering match, down onto the pan. Within the pan would be a small amount of powder which once ignited sent a spark down the vent to set off the main charge in the breech of the barrel.

Despite improvements, the musket remained a very slow and cumbersome weapon to operate, and so to provide a steady barrage of fire, a complete sequence of loading and firing was introduced. The front rank musketeer would fire his musket and then walk back along the file to the rear rank to reload. The next musketeer – having reloaded – would fire and then return to the rear followed by the next and so on, thereby maintaining a continuous cycle of fire.

were clearly spelled out. After this, the first month's pay of three *escudos* per man was handed over to the captain and distributed by him; for many, it would be the last money they would see for months to come.

Money, in fact, was a great problem in the Spanish army; even with silver and gold from America coming into the treasury, there was not enough to pay the soldiers on time. When money did arrive, it was subject to stoppages by the captain for new weapons and other costs, and in the seventeenth century the soldier's pay was quite inadequate in the first place. The monthly wage of three *escudos* did not change for a century, although food prices increased four times in the same period. Arquebusiers and musketeers were paid slightly more than pikemen, but they had to buy their own powder and shot. Not surprisingly, then, the soldier did not rely on his wages, and when the money came – sometimes in quite large sums to cover many months of arrears – he would often spend it all in a few days of drinking, whoring and gambling. The rest of the time he lived on payments in kind and on booty, which could be considerable. Alonso de Contreras boasted after one campaign, 'We returned so rich that I, a private soldier whose pay was only three *escudos*, even I brought back three hundred *escudos* for my part, in clothing and silver. Moreover . . . the viceroy ordered that we should be awarded a further share of the booty and I got my hat crammed to the brim with double silver *reals*.' Nevertheless, he managed to get through this small fortune during three days of debauchery.

Even without regular pay the soldier rarely starved, since he continued to receive the bread ration of $1\frac{1}{2}$ pounds per day. The bread was normally distributed in three-pound loaves every second day, and was, in theory, made of one-third rye and two-thirds wheat. If the bread was good, this was an adequate ration – civilians on poor relief in Antwerp only received a pound per day – but sometimes it was not good, and occasionally it was worse than nothing at all. One ration distributed in 1630 consisted of 'bread' made from offal, unmilled flour, broken biscuits and lumps of plaster, which caused many deaths, and a pestilence which prevented the army from operating in the Netherlands in the following year was attributed to other supplies of 'coarse, black, bad bread'.

Clothing for the troops was supposed to be supplied once a year, and came in two sizes – large and small. It generally followed the fashions of the time, with stuffed and padded doublets and breeches in the late sixteenth century and softer styles as the seventeenth century progressed. Spanish troops, however, tended to wear the white neck-ruff even after it had gone out of fashion elsewhere, and soldiers generally adorned their clothes with silver lace if they could afford it. There was, however, no uniformity of dress; each contractor sup-

plied clothes of the colour and style he thought fit. The troops were distinguished from their enemies by the red saltire crosses on back and breast – a badge taken over from the Burgundians – and later by a red sash and red feathers in the hat. Uniformity was not encouraged until well into the seventeenth century; as an observer wrote in 1610, 'never was there a strict ruling on the costume and armament of the Spanish infantry, for it was this that raised the morale and dash that must possess the men of war.' Another commentator agreed that 'it is the finery, the plumes and the bright colours which give spirit and strength to a soldier so that he can with furious resolution overcome any difficulty or accomplish any valorous exploit'. Nevertheless, one tercio of the 1580s was

dressed all in black and thus earned the nickname 'The Tercio of the Sextons', while another, bedecked with 'plumes, finery and bright colours' was labelled 'The Tercio of the Dandies'.

Officers, too, took pride in their dress, and even if serving with the ranks would try to be accompanied by one or two servants, despite the cost. After his promotion to captain, Alonso de Contreras, described his turn-out at a general review: 'Poor as I was, I had my livery worn by two trumpeters and four lackeys. They were dressed in scarlet uniforms faced with silver, with cross-belts and plumes, and over their uniforms capes of the same colour. My horses – five in all – had their saddles trimmed with silver braid, with pistols showing at their

saddle bows.' Drummers and pipers, too, wore slashed doublets and fancy trimmings. But regardless of how splendid the officers and men looked on the parade ground at the beginning of a campaign, the problems of re-supply ensured that many of them degenerated into 'old ragged rogues' in the course of the following winter.

To help keep out the cold, soldiers would construct huts (barracas, from which we derive the word barracks) out of wood from abandoned cottages, carts, and so on. They were simply made, and normally of a sufficient size to accommodate four men; when the army began moving again, they would always be burned down. When troops were quartered in towns, special barracks in the modern sense of the word would sometimes be built; otherwise the men would have to be billeted in the inhabitants' homes. The largest and wealthiest houses were usually exempt from billeting, which meant that up to five men would have to be housed in even the smallest one-storey hovels.

But it was not only the men who had to be housed. Although married men were not recruited into the army, there was no effective prohibition on marriage once a soldier had joined up. The government did not like it, since married men were considered to be less courageous and more likely to mutiny in protest at the harsh conditions of life in camp, and in 1632 it was decreed that only one man in six should be allowed to marry; needless to say, this restriction could not be enforced. The soldier's wife would work as a seamstress, cook or scullion to help support her family. A veteran of the Netherlands campaign reported in 1623 that most soldiers there were 'married and burdened with children, and with such troops one can achieve nothing because the women promote mutinies, inciting their husbands to disobedience on account of the necessities they suffer and see their children suffering'. Not surprisingly, the authorities made provision for men who chose not to marry, and there were between three and eight prostitutes attached to each company. They were compelled to serve in the guise of washerwomen, and the authorities tried to make sure that they were 'of a competent age' and in good health and stamina. In 1629, the Walloon tercio counted nearly three hundred women to just over 1000 men, but the number of wives and prostitutes was usually considerably lower than this.

The military authorities also made provision for the health of the troops; a permanent military hospital was set up in the Netherlands at Mechelen in 1585, and field hospitals were established during sieges and campaigns. Treatment in the hospitals was free, although they were financed in part by stoppages from pay. The most dangerous wounds were those caused by bullets, since these were quite likely to shatter bones and cause internal bleeding, but light wounds inflicted by sword-cuts and pike-thrusts were relatively easy to cure. In addition, the Mechelen hospital had to cope with large numbers of men suffering from *el mal galico* ('the French disease'), despite the vetting of prostitutes.

There was also a home for wounded veterans, but on the whole the government made little provision for old soldiers. Only a few privileged nobles were made wardens of castles or given similar minor official posts, and the resulting resentment was described by the Marquis of Aytona: 'Some return indignant to their lands, trusting to a relative, resolved to put up with the most abject poverty rather than re-enlist; others go back in despair to the army, and not only do they become trouble-makers in the ranks . . . but many go off and take sides with our enemies.'

An attack on Sienna. The importance of hand guns is made absolutely clear, although they were hardly as effective or easy to use as shown here.

HAERLEM.

A Dutch engraving of the atrocities perpetrated by the Spanish army after the capture of Harlem in 1573, soon after the beginning of the Netherlands war: mass executions blessed by priests while the fife and drum play.

For the force that called itself the 'Catholic Army', there was, of course, provision for the soldiers' spiritual welfare. Besides the chaplain-major of the tercio and his two assistants, each company had its own priest. The piety, or possibly superstitions of Spanish soldiers is shown by the large number of images and crucifixes which they carried, and by the numerous grants to religious confraternities and monasteries in their wills. Some of the grants may not have been entirely voluntary, since wills were usually dictated to the chaplains by illiterate soldiers, but fear of death and final judgement usually brought about a real, if temporary, amendment of character. For the rest of the time, the Spanish rank and file was notorious for the crimes of rape, robbery, murder and arson, especially when officers were absent. The soldiers' religion was one of pious exercises unaccompanied by ethical standards, especially where heretics were concerned; Spanish soldiers were creatures of their age, and the sectarian nature of war in that age meant that neither Catholics nor Protestants paid much attention to the relationship between the soldier's trade and the Christian way of life.

Discipline in the company was controlled entirely by the captain, who could assign men to light or onerous tasks as he chose and could flog or fine a soldier at will and without appeal. Promotion within the company, too, was his prerogative, and a man promoted to corporal immediately earned twice as much as a private soldier. The captain also had a special fund for bonuses which he could distribute as he saw fit. These powers, coupled with the captain's responsibility for stoppages from the men's pay (when it arrived), meant that accusations of favouritism were frequent, and often led to outright mutiny. One of the demands of the mutineers at Antwerp in 1574 was that men should not receive 'a dishonourable punishment, like strokes of the lash, if the offence does not merit it'.

Mutiny was indeed the bane of the Spanish army, especially in the Netherlands, mainly due to the constant and prolonged delays in making good arrears of pay. Between 1572 and 1607 there was a total of forty-five mutinies involving a hundred men or more, some of them lasting over a year. In August 1602 almost half of one army in the Low Countries – more than 3000 men – rebelled and gathered together at the town of Hoogstraten, which they proclaimed an independent republic, and they dressed themselves in green to proclaim their neutrality in the Dutch war; this particular outbreak lasted for nearly three years. Groups of mutineers often devised their own banners, seals and mottoes, elected leaders known as *electos*, and published their demands in print. A contemporary described the rough-and-ready democracy of the mutineers:

90

Decisions are taken by a show of hands. The quarters of the *electo* overlook the square, and from a window he makes his proposals to the squadron. When they are fed up with him they pass from words to bullets. For this reason the *electo* has always a sentinel to watch over him. He cannot receive or transmit correspondence without discipline, for they impale on their pikes or shoot down anyone who commits an offence. Most of their rules are savage, therefore, though some are just and legitimate.

Apart from lack of pay and ill-treatment by officers, mutinies were caused by reaction to the new conditions of warfare: winter sieges were common, and the troops no longer automatically went into 'winter quarters' when cold weather came. A group in 1574 demanded that 'His Excellency [the Duke of Alva] should not keep the soldiers more than six months in the field unless there is great necessity, on account of the great sufferings caused by the frosts and the cold which have caused many soldiers to die, frozen to death on the open road or keeping watch in the trenches'. Instead, 'during the time when the enemy is not campaigning or is not besieging places, Your Excellency should order us to winter-quarters in walled and populous towns'.

Outbreaks of mutiny grew more frequent when the soldiers discovered how successful they could be. Only once during the Dutch war, in 1594, did the authorities try to put down a mutiny by force, and the two attacks by loyalist forces were beaten off. In most other cases, the government was forced to settle with the mutineers by immediate payment of all or a large part of the soldiers' arrears. The money would be brought into the mutineers' garrison under a heavy guard and taken to a church, where each man would present his upturned hat and count the *escudos* or *florins* as they were dropped in.

It may seem strange that an army which endured a permanent spate of mutinies could be so successful. What was important, however, was that the Spaniards could keep an army in the field at all in the dreadful conditions of the Dutch wars. And the mutinies do throw light on the basic strength of the Spanish Army. The mutineers were demanding better conditions; they did not question the role as a permanent army serving Spain. For the first time in Europe since the Romans, an army was in being which formed a military society accepting that its task was to fight in any war if so ordered, and not as a feudal levy, or as a hired mercenary force.

The tactical success of the Spaniards, especially in the early part of the century, was mainly due to their adaption to the possibilities offered by firearms. There were two weapons in use by 1600: the arquebus and the musket. The arquebus was the more common of the two, although its extreme range was no more than two hundred yards. The musket was introduced by the Spanish army in the middle of the sixteenth century, and had a longer barrel and a straighter stock than the arquebus. It was also heavier and fired a heavier ball which was capable of piercing armour at 180m (200 yds). But it was expensive – it cost as much as a musketeer's

pay for a month – and the government had enough trouble paying the men, let alone buying new equipment. The arquebus continued to be used, therefore, well into the seventeenth century, especially for skirmishing.

The shot were protected and augmented by the pikemen, who, in the Spanish army, preferred weapons of about 3m (13 feet) in length instead of the longer Swiss pike. The proportion of pikemen gradually decreased throughout the period of Spanish military dominance. The cost of arms was apparent among the pikemen as well as the musketeers, for a helmet and corselet (back and breast plates, often accompanied by skirt-like tassets) cost enough to pay for two years' bread rations for one soldier. Accordingly, only half of the pikemen wore armour during the earlier part of the period, a proportion rising to two-thirds during the 1630s.

The Dutch war was primarily one of sieges and plundering punctuated by guerrilla attacks, and there is no great set-piece battle which illustrates the tactics used by the Spaniards. But those tactics had, in any case, been drawn up as early as 1503 by Gonzalo de Córdoba to illustrate his theory of the effectiveness of firearms used in conjunction with field defences, and they did not change appreciably through the century. By the time of the battle of Cerignola, which took place during the Italian campaign in 1503, 'El Gran Capitan' had vastly increased his numbers of arquebusiers from the proportion usually employed before his time. Each man had his bandolier containing twelve separate charges of powder – the 'twelve apostles' – and the men were placed behind a rampart of earth and vine-props. The French were allowed to charge to well within effective short range before the arquebusiers opened fire; at Cerignola not one Frenchman reached the rampart. A general advance of the Spanish soon cleared them completely from the field. Córdoba repeated his success against vastly superior numbers at Garigliano later in the same year, again winning the battle with firearms rather than cavalry. It was only in rare cases that the arquebusiers could rely on a rampart for protection, however; normally this was the main function of the pikemen, especially while the shot were reloading. The mingling of pikemen and arquebusiers, combining offensive and defensive functions, came to be known as the Spanish square, and it dominated the battlefields for more than a century.

Spain's loss of the Netherlands and general decline in power and influence were due in part to the difficult nature of the terrain in the Low Countries and the tactics of the Dutch, but the primary reason seems to have been the vast expense of a protracted war fought with such fragile lines of supply as the 'Spanish road' down the Rhine and the ocean route through the hostile English Channel. The government could not continue to finance the war and still defend itself against its other enemies. As the structure of the state declined, so the army, which depended on forming a successful bureaucratic structure nourished and supported by the central government, declined also. In the end, inflation rather than defeat in the field destroyed the efficiency of the *tercios*.

9
CROMWELL'S IRONSIDES

The English Civil War was fought during a period of intense intellectual discussion about the relationship of the individual to the state and society, and of all three to religion; what Coleridge called, 'the grand crisis of morals, religion and government'. The Parliamentary forces had as their spearhead a significant nucleus of men who came to the cause because they were prepared to think about these problems, and act on their conclusions. Of course many enlisted for less worthy reasons: the Parliamentarians paid regularly, at least more regularly than the Royalists; for many it was a way to settle old scores; others were little more than pressed into service. The power on the battlefield of Oliver Cromwell's Ironsides, however, grew out of their reasoned convictions rather than from fear, ambition or coercion; and this makes them an unusual set of fighting men.

When war broke out in August 1642 between King Charles I and Parliament, there was of course no standing army in existence on which either side could draw. The Elizabethan system of 'trained bands' had staggered on through the first third of the seventeenth century, but little training was actually given to those selected. The bands were supposed to be organized within the counties, but by the 1630s things had been allowed to slide so far that there was a chronic shortage of horsemen and of good bloodstock for them to ride. One commentator at the time complained that the cavalry was out of fashion in England. London was the only city in which training was seriously attempted, and as London was the first haven for the Parliamentary leaders it is not surprising that at the start of the war most of their cavalry came from there. As for the rest, both sides had to rely to a large extent on volunteers.

Towards the end of 1642 it had become obvious that the army fighting on behalf of the English Parliament was not going to be able to crush the Royalist forces overnight, as previously hoped. Oliver Cromwell, then one of the recently appointed captains in the Parliamentary cavalry, gave this explanation to Colonel

John Hampden for the Parliamentarians' failure at the battle of Edgehill:

> Your troopers . . . are most of them old decayed servingmen and tapsters and such kind of fellows. Their [the Royalists'] troopers are gentlemen's sons, younger sons and persons of quality; do you think that the spirits of such base and mean fellows will be ever able to encounter gentlemen that have honour courage and resolution in them? . . . You must get men . . . of a spirit that is likely to go on as far as gentlemen will go, or else I am sure you will be beaten still.

The solution which Cromwell proposed and adopted was the recruitment of 'middling' sort of men, whose

Left *One of the Ironsides, in standard equipment with breast and back plates and 'lobster pot' helmet.*
Right *Oliver Cromwell, who combined novel and clear-cut ideas with a formidable military brain.*

religious convictions or political awareness were developed enough to motivate them to fight for their beliefs. 'If you choose godly, honest men to be captains of horse, honest men will follow them,' argued Cromwell. 'A few honest men are better than numbers.' The fighting men of Cromwell's own unit were the willing guinea-pigs for his innovative ideas, and their experiences between 1643 and 1645 mirrored the shifting military, political and religious climate of the time. 'He had special care to get religious men into his troop,' a contemporary witness remarked. 'These men were of greater understanding than common soldiers and therefore more apprehensive of the importance and consequence of war, and, making not money but that which they took for the public felicity to be their end, they were the more engaged to be valiant.'

Cromwell picked the 'godly' recruits for his cavalry from among the farmers of Huntingdonshire, for which county Cromwell was a member of parliament. They were not tenant farmers or agricultural labourers, but freeholders – men who owned their own land, or the sons of such men. All were volunteers who, according to a contemporary, joined up 'upon a matter of conscience'. Many had good reason to be personally loyal to Cromwell, who had vociferously supported his constituents' opposition to the drainage and enclosure of the common fenlands in the 1620s and 1630s. It was not necessary to 'press' men for the cavalry, although this became more and more frequent in the infantry. In the usual way, the cavalry had a higher social standing within the army, were better paid, and had greater opportunities for foraging (and plunder) than the slower-moving regiments of foot.

Cromwell's original commission had been as a captain, with power to raise a troop of eighty men, but by the beginning of 1643 he had been elevated to the rank of colonel. This, in theory, allowed him to raise an additional five troops of horse, but by 1644 his regiment had grown to double strength – fourteen troops in all, each consisting of between eighty and one hundred men. Five of Cromwell's thirteen troop commanders were his own relatives, and there were also two other captains of note: Robert Swallow's troop was raised solely from funds supplied by the young men and women of Norwich, while Ralph Margery, a low-born man, was the subject of several complaints from the county committee which had originally employed him. Cromwell's replies to these are characteristic:

> It may be it provokes some spirits to see such plain men made captains of horse. It had been well that men of honour and birth had entered into these employments, but why do they not appear? . . . I would rather have a plain russet-coated captain that knows what he fights for, and loves what he knows, than that which you call a gentleman and is nothing else. I honour a gentleman that is so indeed.

The most remarkable aspect of Cromwell's own regi-

The religious element was always to the fore in Cromwell's armies; here, his cavalry sing the 117th Psalm after defeating the Scots at Dunbar in 1650.

ment, of course, was the degree of religious commitment shown by the officers and men. As Cromwell's commander-in-chief, the Earl of Manchester, complained in 1644:

> Colonel Cromwell in the raising of his regiment makes choice not from such as were soldiers or men of estate, but such as were common men, poor and of mean parentage, only he would give them the title of 'godly, precious men'. . . I have heard him often times say that it must not be soldiers nor Scots that must do this work, but it must be the godly. . . . If you look upon his own regiment of Horse, see what a swarm there is there of those that call themselves godly; some of them profess that they have seen visions and had revelations.

Cromwell himself, in replying to other accusations of extremism among his soldiers, claimed that 'they are no Anabaptists but honest, sober Christians,' although this was obviously a matter of opinion and definition rather than objective fact. Cromwell, however, believed in religious toleration within the wide spectrum of Puritanism. When the Scots joined as Parliament's allies against the king, it was agreed that all should sign the Solemn League and Covenant which promised that the English would adopt the strict Calvinist and hierarchical tenets of the Scottish Presbyterian church. This led to friction among the more radical English sects who believed in freedom of conscience and refused to be bound by Presbyterian rules. Cromwell did not at this time think that religious difference need interfere with good soldiering and he protected those with unorthodox beliefs. He remonstrated with a Scottish major-general, who cashiered a colonel for being an 'Anabaptist,' a term which was often applied to anyone deviating from the accepted norms of belief. 'Admit he be, shall that render him incapable to serve the public?', asked Cromwell. 'The state in choosing men takes no notice of their opinions; if they be willing faithfully to serve them, that satisfies.' This was a novel principle in the mid-seventeenth century, and one which Cromwell himself found difficult to stand by in later years. He would not, for example, tolerate the most extreme of all, the Antinomians, who held that their possession of the Holy Spirit freed them from the need to observe any law made by men; obviously, no army could function with such men as these in its midst. 'If Noll Cromwell should hear any soldier speak but such a word he would cleave his crown,' observed a colleague.

Each regiment had its own chaplain, chosen by the colonel and eligible for pay of eight shillings per day. Although the chaplain's duties apparently included the writing of battle reports and histories of campaigns, much of their preaching seems to have been taken up with proving the righteousness of the Parliamentarian cause and the wickedness of the enemy. The minister who wrote 'The Souldier's Catechism', a pamphlet produced in 1644, claimed: 'We take up arms against the enemies of Jesus Christ, who in his Majesty's name make war against the Church and people of God,' and called for 'the pulling down of Babylon, and rewarding her as she hath served us'. The victories that the Parliamentarians had obtained this far were a sign that 'Almighty God declares himself a friend to our Party,' and the author assured his readers that 'God now calls upon us to avenge the blood of his saints that hath been shed in the land'. At the battle of Edgehill in 1642 the chaplains in the Parliamentary army, according to an eye-witness, 'rode up and down the army through the thickest dangers, and in much personal hazard, most faithfully and courageously exhorting and encouraging the soldiers to fight valiantly and not to fly, but now, if ever, to stand to it and fight for their religion and laws'.

Preaching, however, was not left to the chaplains alone. There were many cases in which sermons were interrupted by soldiers, and sometimes the preachers were ordered out of their pulpits and replaced by officers or troopers who would then proceed to expound their own views, denouncing the regular minister as a false prophet. Unlicensed preaching was theoretically prohibited within the army, but this regulation was ignored, sometimes with the excuse that there were not enough ordained clergy to preach the word of God adequately. Other spokesmen were blunter, as one minister recorded:

> So soon as I came out of the pulpit, at the very foot of the pulpit stood a man . . . these were his words: 'Sir, you speak against the preaching of soldiers in the army; but I assure you, if they have not leave to preach, they will not fight; and if they fight not, we must all fly the land and be gone . . . these men who are preachers, both commanders and troopers, are the men whom God hath blessed so within this few months, to rout the enemy twice in the field, and to take in many garrisons of castles and towns . . . and I thought good to let you understand so much.'

Cromwell himself defended the right of soldiers to preach, 'which I think has been one of the blessings upon them to the carrying on of the great work', and this contributed greatly to the growth of sectarianism within the army during the Civil War years.

The recruitment of men who could think for themselves, and were encouraged to do so, led to serious problems immediately after the war, when political discussion in the army led to its adopting many ideas (such as those of the Levellers and the Diggers) which were at odds with the professed policy of the Parliamentarian leaders; but the vigorous expression of opinions certainly did not detract from the efficiency of Cromwell's men as fighting men before their final victories over the Royalist armies. Quite the contrary; men who knew what issues they were fighting for were resolute in battle and prepared to undergo the discipline necessary for success.

The Ironsides were a unique force in their religious

The armour of an Ironside. The head, neck and torso were well protected against sword or pike thrusts, but not against bullets fired at short range.

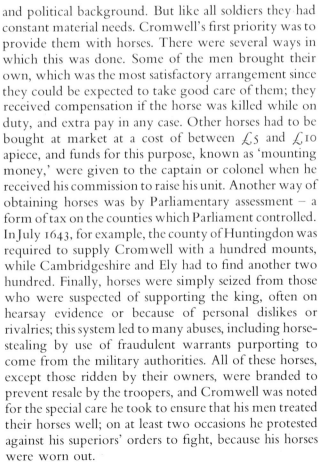

and political background. But like all soldiers they had constant material needs. Cromwell's first priority was to provide them with horses. There were several ways in which this was done. Some of the men brought their own, which was the most satisfactory arrangement since they could be expected to take good care of them; they received compensation if the horse was killed while on duty, and extra pay in any case. Other horses had to be bought at market at a cost of between £5 and £10 apiece, and funds for this purpose, known as 'mounting money,' were given to the captain or colonel when he received his commission to raise his unit. Another way of obtaining horses was by Parliamentary assessment – a form of tax on the counties which Parliament controlled. In July 1643, for example, the county of Huntingdon was required to supply Cromwell with a hundred mounts, while Cambridgeshire and Ely had to find another two hundred. Finally, horses were simply seized from those who were suspected of supporting the king, often on hearsay evidence or because of personal dislikes or rivalries; this system led to many abuses, including horse-stealing by use of fraudulent warrants purporting to come from the military authorities. All of these horses, except those ridden by their owners, were branded to prevent resale by the troopers, and Cromwell was noted for the special care he took to ensure that his men treated their horses well; on at least two occasions he protested against his superiors' orders to fight, because his horses were worn out.

Cromwell's own cavalry were 'shot on horseback' (cavalry armed with firearms) of the type known as 'harquebusiers'. The term was something of a misnomer, however, as the harquebus or carbine had been largely abandoned by this time. Troopers wore steel back- and breast-plates over coats of buff leather, together with light helmets open in front (known as the 'pot'), breeches and boots. They carried swords and sometimes also small pole-axes; but their only firearms were pistols. The nickname of 'Ironsides' which Cromwell's regiment earned was a reflection of their steadiness in action rather than the quantity of their armour. It was said of the Ironsides that 'being well armed within by the satisfaction of their conscience, and without by good iron arms, they would as one man stand firmly and charge desperately'. The back- and breast-plates and helmets were lined with leather and cost about thirty shillings a set, while the sword cost about five or six shillings; the wheel-lock pistols were complicated to make, and might cost from twenty-six to thirty-eight shillings a pair. The saddle, finally, was a substantial one, and cost about sixteen shillings. There is some evidence that the officers, unlike the troopers, carried carbines in addition to their pistols and swords. The pay of a trooper was, in theory, two shillings per day, to be paid out every twenty-eight days; from this he was expected to provide food for himself and his horse, together with the cost of his lodgings, clothings, and horseshoes. By comparison, officers were well paid, although figures for the first Civil War are difficult to assess, as they fluctuated wildly from place to place. From 1648, however, a cavalry colonel's daily pay was twenty-two shillings, a major's fifteen shillings and eightpence, a captain's ten shillings, and a lieutenant's five shillings and fourpence. At this time the ordinary soldier's pay was still only two shillings and threepence, so it is safe to assume that officers' pay in the earlier period was much the same, at least on paper, as it was after 1648.

The regularity of a soldier's pay depended on the ability and willingness of the county committees to provide it through assessments on its inhabitants. In the case of Cromwell's Huntingdonshire, a poor county, the pay was usually in arrears (the troops were better off in this respect than the Royalist forces, however), although Parliament itself occasionally rewarded him with *ex gratia* payments after victories over the enemy; when his personal resources were exhausted, Cromwell was forced to send begging letters to other counties and to influential friends, pleading that 'the heavy necessities my troops are in press me beyond measure'. The pay of the forces of the Eastern Association was placed on a sounder footing in 1644 through the setting up of a central treasury, after which the men were paid more regularly.

Since the armies in England did not use tents, the troops were usually billeted on the local inhabitants. In general, the individual soldier did not pay for his lodging in cash; instead, his officer would give the landlord a receipt which had to be presented at the county town for payment, and the sum was then deducted from the county's contribution to the army. The cost of a day's food and lodging for a trooper and his horse was about one shilling and fourpence, leaving him with eightpence for clothing, horseshoes and other needs.

During a campaign the men might sleep in barns or in the open fields, but their food still had to be provided for. The only provisions supplied in most cases were bread and cheese, which could be carried in the soldier's own 'snapsack'; his drink was beer, the common beverage of every household at that time. The Parliamentarians relied on the goodwill of the country people for supplies, a goodwill which was not always very evident. The citizens of London, however, were glad to supply Essex's forces at Turnham Green where 'there were sent at least a hundred loads of all manner of good provisions and victuals, bottles of wine and barrels of beer . . . accompanied by honest and religious gentlemen; who went to see it faithfully distributed to them'. The standard ration was normally one pound (0.45kg) of bread and half a pound (0.22kg) of cheese per day, but

this was often reduced to little or nothing: one account of the battle of Marston Moor shows that the soldiers fought on little but faith: 'Through the scarcity of accommodations very few of the common soldiers did eat above the quantity of a penny loaf, from Tuesday till Saturday morning; and had no beer at all.' As usual, the troops had to supplement their diet with whatever they could buy or steal; the theft of cattle and sheep was especially common in the early days of the war, when discipline was still very slack. Soldiers, like other Englishmen of the seventeenth century, ate few vegetables.

The discipline which Cromwell maintained in his own regiment was one of its most striking characteristics: 'No man swears but he pays his twelve pence; if he be drunk he is set in the stocks or worse; if one calls the other "Roundhead" he is cashiered; insomuch that the countries where they come leap for joy of them, and come in and join with them.' The laws and penalties which were observed throughout the army were the familiar ones of death by firing squad for mutiny, whipping or riding the wooden horse for theft from country people or from protected prisoners of war, and so forth. Desertion was in general so common that the authorities were usually unable to punish it; all they could do was ask that those who had 'gone to visit their friends' should return to their posts as soon as possible. In Parliament's army, blasphemy was sometimes punished by boring through the tongue with a red-hot iron; on the other hand, the religious attitudes of many Puritans positively

The bureaucracy of 17th century war: in this case a receipt for ordnance stores and tools used by the armourers of an artillery train, early in 1644.

Civil War Cavalry

During the early stages of the English Civil War there were two broad schools of cavalry tactics. The Royalists favoured charging at the gallop relying on shock power to overwhelm the enemy, while most Parliamentaries adopted what was known as the Dutch fashion which emphasised firepower. The initial success of the Royalist cavalry convinced Oliver Cromwell that Parliamentarian tactics were unsuitable, and he ordered his Ironsides to adopt the Royalist approach but in a more ordered manner. Well equipped and armed, and well trained, the disciplined Ironsides were the elite of the Parliamentarian army and proved a match for the cavalry of Prince Rupert.

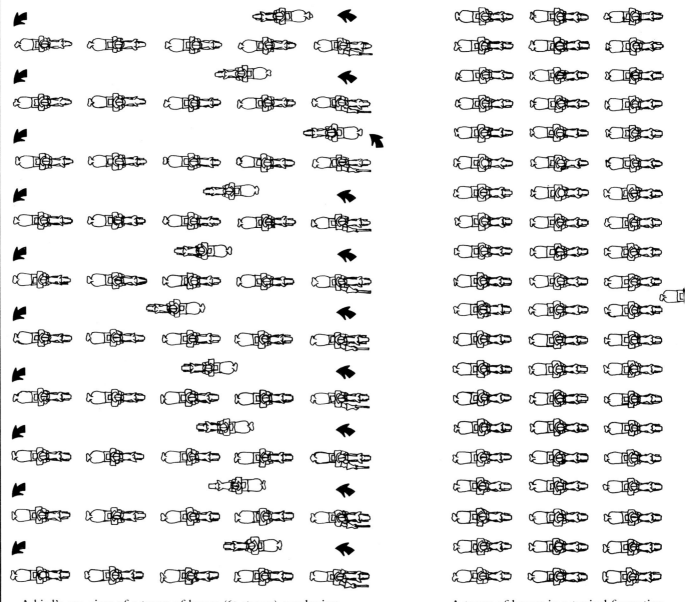

A bird's eye view of a troop of horses (60 strong) employing the Dutch school of tactics. Derived from the manoeuvres of the German Reiters, the troop would trot up to the enemy and the front rank would fire its pistols (usually two, sometimes four) and trot back down the files to become the rear rank, so allowing the next rank to loose-off its pistols. After this barrage of pistol fire, the enemy — in theory at least — would be sufficiently weakened to allow the cavalry to charge home with the sword.

A troop of horses in a typical formation used by the Ironsides. Marshalled into three ranks they would advance against the enemy at a fast trot and use the shock effect of sword and horse to gain success.

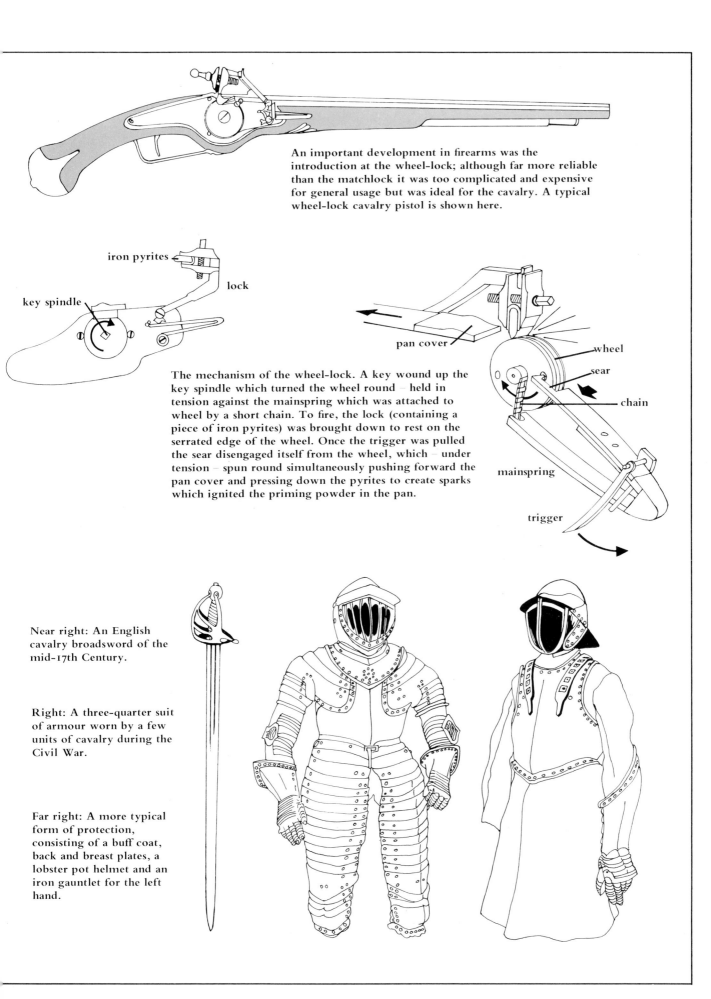

An important development in firearms was the introduction at the wheel-lock; although far more reliable than the matchlock it was too complicated and expensive for general usage but was ideal for the cavalry. A typical wheel-lock cavalry pistol is shown here.

iron pyrites

lock

key spindle

pan cover

wheel

sear

chain

The mechanism of the wheel-lock. A key wound up the key spindle which turned the wheel round – held in tension against the mainspring which was attached to wheel by a short chain. To fire, the lock (containing a piece of iron pyrites) was brought down to rest on the serrated edge of the wheel. Once the trigger was pulled the sear disengaged itself from the wheel, which – under tension – spun round simultaneously pushing forward the pan cover and pressing down the pyrites to create sparks which ignited the priming powder in the pan.

mainspring

trigger

Near right: An English cavalry broadsword of the mid-17th Century.

Right: A three-quarter suit of armour worn by a few units of cavalry during the Civil War.

Far right: A more typical form of protection, consisting of a buff coat, back and breast plates, a lobster pot helmet and an iron gauntlet for the left hand.

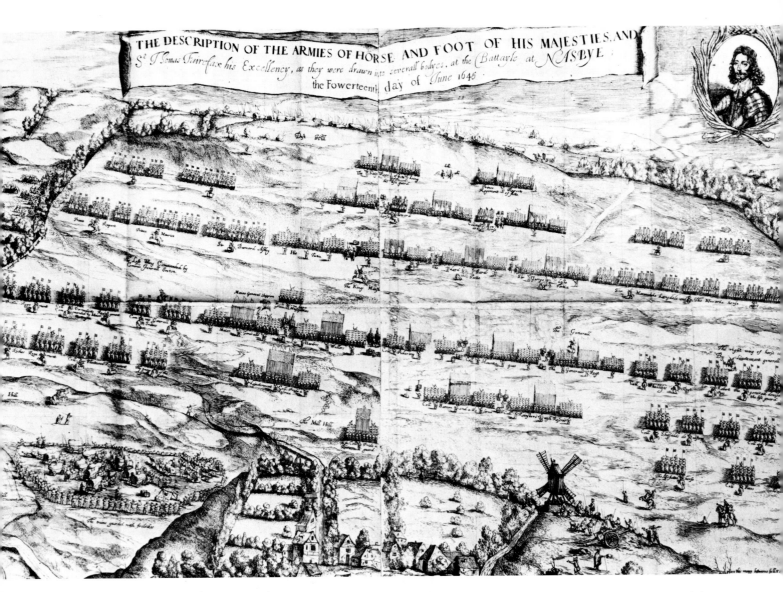

The battle of Naseby was the crowning victory of the Civil War, and Parliamentarian success owed much to the discipline of the Ironsides stationed on the right flank.

encouraged the desecration of church buildings and the destruction of stained glass and monuments.

In 1642, when the war began, Cromwell had had no previous military experience whatsoever. Many of the other officers however, had served in the Low Countries or Germany in the 1620s and 1630s. Cromwell learned quickly, adopting the best and most practical of his opponents' tactics and using the strong discipline of his unit to full advantage. The first engagement in which he commanded the Parliamentarian forces was at Grantham in Lincolnshire in April 1643, and for the first half-hour both Cromwell and his Royalist opponents tried to make up their minds when to attack. Finally, Cromwell records:

> They not advancing towards us, we agreed to charge them . . . after many shots on both sides, we came on with our troops at a pretty round trot, they standing firm to receive us; and our men charging fiercely upon them by God's Providence they were immediately routed, and all ran away.

Having learned the importance of taking the initiative in cavalry engagements, Cromwell refused to stand still and await an oncoming Royalist advance at Gainsborough, three months later, even though his men were not properly formed up:

> In such order as we were, we charged their great body . . . we came up horse to horse, where we disputed it with our swords and pistols a pretty time, all keeping close order, so that one could not break the other. At last they a little shrinking, our men . . . pressed in upon them, and immediately routed this whole body . . . and our men pursuing them, had chase and execution about five or six miles.

That was not the end of it, however, for the Royalists had not committed their reserve; Cromwell had been careful to keep three companies of his regiment from joining in the pursuit, and these succeeded in driving the remaining Royalists into a quagmire and killing their general. This second victory testified to the strength of the discipline exercised by Cromwell over his troops of horse, and the thorough training which allowed them to advance knee to knee without breaking.

The seventeenth century battlefield was fought over by troops using a combination of arms. Infantry might use pikes or primitive firearms, while cavalry tactics could take the form of an outright galloping charge, the controlled, but less ferocious, charge at a trot or a wheeling formation, using firearms. Cromwell's men perfected the steady charge at a trot, ignoring the few casualties caused by enemy fire, and eschewing the heroic gallops of the Royalist cavalry. Prince Rupert's horsemen sometimes carried all before them, but they were difficult to reassemble after early success. Cromwell's men never were. Their steady control and relentless precision made them masters of the battlefield.

Just as the original scheme whereby the regiments were raised within, and paid by, the individual counties had been superseded by the creation of associations of counties with central treasuries, so in 1645 the army was reorganized once again, this time on a national basis. In order to overcome the reluctance of the troops to leave the counties or associations which paid them and to put the regiments' financial support on an equal footing, the New Model Army was formed from an amalgamation of existing regiments plus new units of infantry consisting of pressed men. Cromwell's double-strength regiment of horse was divided between the new formations: six of the troops went to make up the new regiment of Sir Thomas Fairfax, the commander of the New Model Army; six more troops became the regiment of Colonel Edward Whalley, Cromwell's cousin and his former second-in-command; and the remaining two troops provided the nucleus of other units. Thus the formation of the New Model marked the end of the Ironsides as a regiment, although ironically, the six troops of Fairfax's regiment became Cromwell's once again in 1650 when he took over the command of the army on Fairfax's retirement.

No doubt some of the troopers who joined Cromwell in 1643 were still serving in 1650, but others had, of course, been discharged over the years because of wounds or ill health. Cromwell's regiment, like other units, had its own surgeon and his two assistants, but in the absence of military hospitals it was usual for unfit men to be left in their quarters when the regiment marched on. The householder or a local physician would be made responsible for looking after them, and the cost of this was met either from the fortnight's pay which was usually given to each unfit soldier when the regiment moved on, or else from the issuing of receipts redeemable at the county headquarters, as with ordinary payments for quartering. But the money often ran out or was unavailable because of the distance to the headquarters, and a number of heart-rending petitions from impoverished householders survive. After the battle of Edgehill, one Hester Whyte nursed some wounded men, 'who continued at her house in great misery by reason of their wounds for three months. She often sat up night and day with them, and, in respect of her tenderness to

Sir Thomas Fairfax, the first commander of the New Model Army which incorporated the Ironsides, and ended their existence as a semi-independent force.

the Parliament's friends, laid out her own money in supply of their wants.' It seems probable that many sick or wounded soldiers, as soon as they had recovered sufficiently to travel, made their way home without leave and put all thought of army life behind them.

The fate of men from disbanded units was little better; lack of money in the treasury meant that their arrears could rarely be paid in cash. The certificates of arrears which they received instead were soon regarded as worthless. For their future livelihood, the men disbanded during the 1640s had, in general, only agricultural work, begging or highway robbery to choose from, and there was a notable increase in the last – although many robbers were former Royalists, who were in a worse plight than their former enemies. For the Parliamentarians, the situation improved during the 1650s as the state eased the restrictions on the practise of trades in towns for old soldiers, and this continued after the Restoration of the monarchy in 1660; two years later the 'middling men' of Cromwell's army had returned to their roots. The diarist Samuel Pepys recorded then that 'of all the old army, you cannot see a man begging about the streets. . . . You shall have this captain turned shoemaker; the lieutenant a baker; this a brewer; that a haberdasher; this common soldier a porter, and every man in his apron and frock, etc., as if they had never done anything else'.

10
THE ARMY OF FREDERICK THE GREAT

When Frederick II came to the throne of Prussia in 1740, he inherited a relatively new kingdom whose military administration was of great efficiency within the eighteenth-century European context. Frederick's achievement was to take the institutions created by his immediate predecessors and build them into the most successful military machine of the age. Frederick's use and development of his father's army astounded all Europe. What the Prussian army seemed to demonstrate was that sheer drill and discipline could turn soldiers into the perfect fighting men; that mere automatons were the ideal men to fill the ranks.

Prussian victories under Frederick were masterpieces of eighteenth-century tactics, in which the superior drill and manoeuvrability of Frederick's troops enabled the king to develop certain methods of attack – notably the oblique order, in which the Prussian army concentrated its efforts against one flank of the enemy while refusing contact on the other. From his attack on Silesia in 1740 to his death in 1786, Frederick led his country through two major wars (1740–48 and 1756–63) in which he first seized territory from the Hapsburgs and then defended it against a coalition of the greatest European powers – France, Russia and Austria.

Frederick manipulated the whole economy of Prussia in the interests of military supremacy, paying lip service to the Enlightenment while in practice developing the machinery of state and the army in a ruthlessly autocratic way. He built upon foundations laid by his father, Frederick William I, a debt he freely acknowledged:

> Only his care, his untiring work, his scrupulously just policies, his great and admirable thriftiness and the strict discipline he introduced into the army which he himself created, made possible the achievements I have so far accomplished.

Compared to Britain or France during the same period, Prussia's was a backward, essentially peasant economy, territorially large but underpopulated. By the end of Frederick's reign, the population had grown to

A Prussian officer and grenadier, wearing the blue coats of the Prussian army. The mitre cap was common to the grenadiers of almost all 18th century armies.

about 8,500,000, composed of nobles and peasants, landlords and tenants – and officers and men. In the Prussia of Frederick the Great, these terms all described the same two groups, for the army could call on the service of almost every peasant and journeyman in the state, although it also relied heavily on foreign recruits. In 1740, the peace-time army was already 80,000 strong; the Hapsburgs had an army of only 100,000 men in a population ten times the size of Prussia's. During the height of the Seven Years War, the Prussian army accounted for 4.4 per cent of the population, compared to 1.6 per cent in France. The rigid stratification of society was actively encouraged by Frederick because it enabled him to organize the whole of Prussia into maintenance of the army; when a peasant was not actually under arms he worked the fields to pay the heavy taxes which supported those that were. One modern German critic has, with some justification, described Frederick's economy as 'backward, retarded, autocratic and sergeant-like'. Duty and service were indeed the cornerstones of his regime, virtues which he personally practised, but which also created the popular myth of the upright, humourless, Prussian officer who sacrificed his individuality for the good of the state.

As with all myths, there is some truth in it. Major-General Joseph Yorke, who reported on the Prussian army in 1758, noticed this trait. 'The service is certainly done with exactness,' he wrote, 'but with less life and gaiety than anywhere I have yet seen.' This was as true of the officer class as it was of the men, since officers, almost always members of the rural nobility, were trained from boyhood for their allotted role.

Officers were mainly recruited from the Junker class of country gentry, who were efficient, hard-working and usually had a modicum of education. They exerted an almost feudal power over the peasants who worked their estates and who provided the bulk of the fighting men. In the battlefield, as in the harvest-field, the Junker commanded and the men obeyed. Quite naturally, the prevailing belief amongst the Junkers was that they enjoyed an inherent superiority which gave them the right to be obeyed. Of course, there was a middle-class, bourgeois or professional element in Prussia, made up mainly of merchants and bureaucratic officials as well as

the nobility and peasantry; and many members of this urban bourgeoisie enjoyed a degree of independence from the semi-feudal society of the countryside. In fact in the latter part of Frederick II's reign, two-thirds of the million and a half exemptions from military service were citizens of towns exercising their traditional privileges. Prussia's rulers up to the time of Frederick the Great had an inconsistent attitude concerning the place of this bourgeois class in the army. Before the time of Frederick's great-grandfather, the German infantry appears to have drawn its officers strictly from the ranks of the nobility. From 1640 to 1688 under the Great Elector, this restrictive practice was relaxed, and both nobles and bourgeoisie were employed as officers. Later, under Frederick I (1688–1713), there was a tendency to rely on the nobility, although he appeared to distrust the old Junker families and allowed some social mobility in both his bureaucracy and in the army. There were instances of bourgeois appointments among the officers; the king

even took the trouble to assure the middle class officers that there would be no discrimination against them when it came to promotion. Nevertheless, the nobles certainly regarded their bourgeois colleagues as inferior, partly because many of them were called upon to serve in the artillery, where their technical skills were needed but which conferred on them an inferior status – almost that of an artisan.

During the reign of Frederick William I (1713–40), the nobility came to predominate overwhelmingly. This was partly the result of economic changes, for the estates of the nobility could no longer support their sons, and the army was an acceptable profession for them. But it was also a deliberate policy of the king, who ruthlessly engineered the composition of his officer corps to his own design. As his son Frederick the Great later recorded: 'In every regiment the officers corps was purged of those people whose conduct or birth did not correspond with the profession of men of honour which

104

Typical uniforms of the guard units of Frederick's army. The two officers in the centre are distinguished by the sashes worn around the waist.

forced him to refill the officer ranks from the bourgeoisie, only to sift them out again after the fighting had stopped. 'The breed of the old Prussian nobility is so good,' he claimed, 'that it deserves to be preserved.'

In any case, there was some laxity in the definition of a nobleman. As in most non-British contexts, it included not only the peerage but also the *nobilitas minor*, that is, all who were entitled to coats of arms, including knights, esquires and gentlemen. In Prussia it was possible to raise a man instantly to the ranks of the nobility merely by inserting the style 'von' in front of his surname and thus granting him a title. One or two plausible aspirants managed to do it themselves and went undetected.

Whatever their social class, men could enter the Prussian officer corps by a variety of routes. The most important single route, however, accounting for about a third of all the officers, was through the Prussian Cadet School in Berlin. During the reign of Frederick William I, the school had been developed into the archetypal tough, harsh institution for moulding the character of adolescent boys. Each noble family was obliged to contribute one son, who at the age of about thirteen was escorted away to the Cadet School under armed guard. Once there, the new cadets were officially under the control of officers, NCOs, and civilian teachers, but the dominating force in their lives was undoubtedly provided by the senior cadets, who traditionally bullied and terrorized the smaller and weaker boys.

Frederick the Great introduced some modifications in his father's harsh régime: he insisted that the cadets should be treated with the respect due to young noblemen, and he also improved the academic side of their education. In 1759 he appointed a new and more humane governor who introduced better domestic conditions, including the provision of a single bed for each boy.

Once he emerged from the Cadet School, the new officer joined his regiment, and there was, of course, keen competition for places in the better regiments. Thereafter, as in the modern army, his chief loyalty would be to his unit rather than to the army as a whole; in the case of Frederick's Prussia, however, it was the company rather than the regiment which claimed his particular affection. If a company captain was promoted, he still retained nominal control of his old company and delegated its day-to-day running to a lieutenant. Joseph Yorke noted that captains were obliged to provide their subalterns with their keep so that the latter had 'nothing to attend to but their duty; whilst quarrels, caballing and all other inconveniences of too many young men messing together are avoided'.

In time of war the officer faced the normal hazards along with his men, but in peace-time there were compensations. He was well dressed in uniform and well accommodated on manoeuvres and campaigns; once he reached the rank of captain he was also comparatively well paid. It was then that a captain could solve his financial problems by pocketing (quite legally) the money which the state advanced to provide for the needs

they sought to follow.' The result was that by 1759 very few of the field officers came from outside the nobility, while all thirty-four of the officers of the rank of major-general and above were noble.

Frederick the Great was himself as capricious in his approach to this question as all three of his predecessors. Frederick preferred the nobility to the bourgeoisie; he claimed to believe that only the hereditary, privileged class had the in-built sense of honour that was needed to lead men into battle. The commercial middle classes on the other hand could too easily be diverted by private interests, and to allow them to reach commissioned rank would be 'the first step toward the decline and fall of the army'. But this was written in 1775; during the Seven Years War (1756–63), however, the army's losses had

106

Left Frederick during the War of Austrian Succession, in 1745. The plain dark blue coat was his way of showing that he was as one with his officer corps.

Above A Potsdam regiment advancing into enemy fire at the battle of Leuthen (1757). Officers would rarely have led their men into battle in such an exposed fashion.

of his company, while sending the men home on leave for several months at a time. The officer's mornings were spent on duty, while his afternoons were largely his own. The younger and less established would probably go off to the local inn to drink wine or beer, or to gamble, or else to visit the local brothel. Senior officers were more likely to put on their best dress uniforms and occupy their leisure time with the local salon society, especially when garrisoned in Berlin. It was hardly the intellectual French salon which Frederick had so much admired as a young man; in fact, many of the officers could not read or write, in spite of the improvements to the academic side of the Cadet School. But Frederick had little time for unthinking officers. He wrote scathingly of the French in 1758: 'Their officers have learnt a military jargon, but they are simply parrots who have learnt to whistle a march and know nothing else.' Frederick himself, in fact, took a great interest in his officers and tried to know each one personally. He regarded them somewhat as his personal servants, with whom he could do as he liked. Depending on Frederick's mood, an officer might find himself a major-general in his mid-thirties, or he might

be stuck in the junior ranks for the duration of his service. If he incurred the king's displeasure, even for a trivial or non-existent offence, he might suffer the humiliation of arrest and possible imprisonment, or even be cashiered. Bourgeois officers came in for particular scorn. When an officer did not measure up to his standard, Frederick's approach was cruel. He would beat an offender in public, even on parade, with his stick, and he hauled one young officer out and dismissed him with the insult: 'Now you can be a clerk, like your father.' On the other hand, Frederick allowed the youngest lieutenant to see him as readily as the most senior general, and there were no badges of rank on the officers' uniforms. Yorke observed that the king 'never gets into a coach but constantly marches on horseback with his infantry, begins his march with them, and leads them into camp or quarters'.

This lack of distinction within the officer corps was an effective way of cementing their unity as a class – the only class that counted, according to Frederick, and one which must not mix with its inferiors. 'The officers must not be permitted to go about with common folk and townspeople,' he said. 'On the contrary, they must seek the company of the higher officers and such of their comrades as behave themselves well and are fired with ambition.' Frederick, too, was the first monarch habitually to wear uniform, and it was not an elaborately braided and glittering commander-in-chief's costume but the ordinary uniform of his officer class. If he wore their coat, then by the same token every officer, from the

in the areas where they lived, and thus became liable for recruitment into their local regiment at any time between the ages of eighteen and forty, each regiment having a designated recruiting area. There were a few categories of exemptions such as merchants, trained apprentices, theological students and widow's sons. Although these exemptions increased throughout Frederick's reign, every other young man could expect to spend a period of about two years in the army. Then, when he was considered fully trained, he was allowed to return home to pursue his peace-time occupation for up to ten months of the year, joining up again for the spring and summer manoeuvres. But even while working in his own fields he was obliged to wear either his uniform or

Above *A grenadier of the 6th Guard Battalion takes aim while his officer gives orders.*
Right *A romantic view of the 18th century battlefield, which nevertheless portrays a true incident, when on 18 June 1757 the 1st Guard Battalion resisted simultaneous attacks from front and rear by Austrian cavalry.*

youngest lieutenant upwards, could feel that he was wearing his king's coat. And although Frederick himself might humiliate one of his officers in public, either verbally or physically, he would never allow anyone else to do so in his presence.

The king's promotion of the dignity of his officer corps extended outside the camp, too. In every sphere of life in Prussia the officer had preference over the mere civilian, no matter how grand, for civilians on the whole would not be members of the nobility and therefore, in Frederick's view, had no independent existence. But even the civilian nobility, diplomats and government ministers had to stand aside at the approach of a Prussian officer. Although the officers were often ignorant, narrow rural nobles, they received enormous social privileges in return for commanding troops in precisely the manner laid down by the king and enduring the caprices of his personality. The ordinary soldiers of the army, in contrast, enjoyed no such social advantage in return for their privations.

Almost every male in Frederick's Prussia was a potential soldier. Male children were registered on rolls

an insignia showing that he was liable for mobilization. As a result, there was virtually no truly civilian element in the agricultural working classes, and the army's attitudes to order and discipline permeated the life of the community at large.

In the circumstances, it is hardly surprising that the recruiting sergeant was one of the most heartily detested figures in the countryside, and there are stories of entire villages taking to the road to avoid him and try to save their menfolk from being ensnared. But there was little hope of evasion, as the recruiting sergeant, list in hand, proceeded with great thoroughness.

The kingdom was divided into cantons, each of which was responsible for a regiment. Infantry regiments were made up of 5000 muskets, cavalry of 1800 horse.

The cream of the young men would be selected for the cavalry; the tall and brave Pomeranians and Brandenburgers were destined to be grenadiers or musketeers in the infantry regiments of the line; lesser breeds like the men of Silesia, Westphalia and Berlin – for Frederick had a low opinion of the capital's residents – went to the less prestigious fusilier units, and were issued with smaller hats and muskets. Hunters, gamekeepers, foresters and the like were a special case; they went to the new *Feldjäger* (light infantry) corps to serve as outlying troops protecting advance reconnaissance parties, to act as guides, and to carry out other duties where their special skills and marksmanship could be of use. Finally, men with an

aptitude for technology might be placed in the artillery. Gunners had a lower status than even the fusiliers, since their job was considered to require merely hard work and precise calculations rather than smartness, bravery and military *élan*.

But Frederick could not maintain the huge army he desired from home recruitment alone. During the Seven Years War he was forced to allow the recruitment of a number of 'free battalions' consisting of foreign mercenaries, although there still had to be a minimum of one third native Prussians in the army at any one time. These free battalions might more accurately be described as roaming bands of brigands, ex-prisoners, deserters, including entire battalions of Austrians, and sundry other undesirables, unreliable in combat and positively dangerous in peace-time. Many of them were prisoners taken in battle and forced to fight against their old comrades; naturally, they had to be used in battle under the constant direct control of an officer to prevent desertion, especially as they were usually employed as cannon fodder in the early battles of a campaign. Frederick sought to preserve his native Prussians by 'using up' the men of the free battalions first. At the end of the Seven Years War those remaining were arrested and disarmed, their officers dismissed, and the men dispersed among the native infantry units.

Once he was enrolled in the army, the newly recruited soldier was fitted out with his basic uniform, sworn in, and assigned to his barracks to begin his basic training. The Prussian practice was for the recruit to spend a great deal of time in the care of a subaltern for individual tuition, often by himself, occasionally as a member of a small group. The subaltern's duty was to teach him the rudiments of drill and weapon handling before sending him to his regular company. In these initial stages of training the recruit was not abused; according to the English observer John Moore:

> [he] is at first treated with gentleness; he is instructed only by words how to walk and how to hold up his head and to carry his firelock. He is not punished, though he should not succeed in his earliest attempts. They allow his natural awkwardness and timidity to wear off by degrees; they seem cautious of confounding him at the beginning or of driving him to despair and they take care not to pour all the terrors of their discipline on his astonished senses at once.

This gentleness was soon replaced, however, by the full horrors of Prussian discipline. The object of training was to reduce the men to puppets, who obeyed unthinkingly and marched unerringly. The rationale behind the Prussian discipline was that, with the crude firearms of the day, men had to be prepared to stand in line to fire volleys at similarly exposed troops or to repel charging cavalry; and the only way to make troops stand in such exposed formations, with death always likely, was to reduce their capacity for independent action to nothing. The troops had no personal reasons for fighting in defence of Frederick's dynastic aggrandisement; all that would motivate them was fear of the penalties they would incur for not obeying orders.

Frederick wrote in 1752, 'unless every man is trained beforehand in peace-time for that which he will have to accomplish in war, one had nothing but people who bear the name of a business without knowing how to practise it'. General John Burgoyne was in the minority among foreign observers when he disagreed with the Prussian method of 'training men like spaniels by the stick'. By the end of Frederick's reign most other armies were emulating the constant drill which Frederick insisted was the only way to impose on the rank and file that 'order, discipline and astonishing precision which made these troops like the works of a watch, the wheels of which by artful gearing produce the exact and regular movement'. The recruit's training concentrated on marching at the correct pace, wheeling and field manoeuvres which would allow his unit to concentrate its fire on the enemy. Frederick knew only too well that with his state's slender resources of manpower and its long borders to protect, he would suffer from a permanent deficiency of numbers compared with any likely enemy, even after he had recruited large numbers of foreigners. The only hope of victory lay in local superiority, the concentration of force against weak spots in the enemy lines. This concentration was possible only if troops could be moved quickly and precisely.

A wide variety of manoeuvres were evolved to achieve this. The simplest was to arrange for the infantry to arrive on the field parallel to the enemy's line, so that each platoon could simply execute a left turn to face the enemy. Should the parallel approach march be impossible, the battalions could arrive in column of open platoons, approaching the enemy on a perpendicular line. Several possible choices then followed to bring the men into line, some of them of stupendous complexity. Each platoon could wheel in turn on reaching the head of the column, then execute a left wheel to face the enemy (processional deployment); or each platoon could make a right turn out of the column, march to its own designated point on the field, then make a left turn to face the enemy (deployment *en tiroir*); or there was an impressive oblique movement of each platoon from its place in column to its place in the line, in which every soldier was required to carry out a cumbersome but spectacular step involving the crossing of one leg in front of the other. Protocol required the senior unit to fight on the right of the line; the wrong troops must not be allowed into that privileged position.

One of the most impressive aspects of the whole drill was the measured pace at which the Prussian infantryman was accustomed to march. He moved at a steady seventy-five paces to the minute, with a recommended stride of 71cm (28 inches). This solemn tread was implanted into the recruit's habit through long and constant practice, and gave the army the air of a group of men unhindered by haste yet firm of purpose. When a specific manoeuvre required extra speed, the infantry changed to the alternative step of 120 paces to the

The imposition of Frederick's will after his invasion of Saxony led to scenes such as these, where magistrates and merchants are being arrested by the army.

minute. As the Scottish observer David Dundas reported after witnessing their drill:

> Although the general movements of the Prussian infantry appear slow and solemn, yet they are so accurate that no time being lost in dressing or correcting distances, they arrive sooner at their object than any others, and at the instant of forming they are in perfect order to make the attack.

Once in place, on the battlefield, the infantry began firing volleys of musket-shot, either by platoons or as a complete battalion. The main weapon that the recruit was given and taught to use was the musket, employed either as a firearm or as a support for the cold steel of the bayonet. The sword was seldom drawn, but was retained largely for sentimental and decorative reasons. Unfortunately, the infantryman of Frederick's army was equipped with one of the least effective muskets in use anywhere in the mid-eighteenth century. It carried a barrel 3 feet 5 inches (1m) long and of $\frac{3}{4}$ inch (2cm) calibre, with a carved stock of walnut (later, in the interests of economy, maple). Its iron ramrod was stronger than the earlier wooden one; on the other hand, it was also heavier. In fact with ramrod and bayonet in place the musket was so unbalanced that the muzzle

dropped downwards, in the hands of all but the strongest soldier, causing most of the shots to fire low. Because of this, musketry was more a corporate than an individual art; the officer in charge of a platoon would try to achieve blanket coverage of the target.

The intention was to bring down the enemy with an overwhelming fire, the ordered ranks producing five or six volleys each minute at the officers' commands. The reality was somewhat different: after a few orderly volleys, even Prussian discipline wavered, and the men would fire as soon as they had reloaded; confused and frightened by the enemy fire bringing down men all round them, they resorted to firing at any target they could see.

If the firepower of the ranks of infantry failed to bring down the enemy or put him to rout, the men were expected to use the bayonet. Frederick had developed unshakeable faith in the power of cold steel, insisting that the bayonet should be fixed at all times when the soldier was on duty, and declaring in 1743 that no enemy would dare stand before his army's bayonets. In fact, however, the bayonet was very rarely used in practice; its effect was more psychological than physical. 'It is a bold front which defeats the enemy, not fire,' he claimed. 'You decide that battle more quickly by marching straight at the enemy than by popping off with your muskets, and the more quickly the action is decided the less men you lose.' He later changed his mind and admitted that the men who loaded faster were likely to get the best of the

111

Eighteenth-Century Firepower

The army of Frederick the Great was the best in Europe mainly because it was the best drilled. Frederick demanded that his army be able to march onto the battlefield in a number of different formations and be able to deliver sustained volleys of fire against the enemy without being disorganized. In order to secure the necessary precision of movement and firepower a ruthless discipline was imposed on the troops. While this reduced the men to little more than automatons, it ensured that they could both inflict and endure murderous point-blank fire.

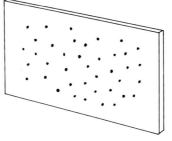

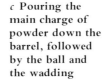

Towards the end of the 17th century the matchlock was replaced by the more efficient flintlock. Except at short range, however, the smoothbore musket remained an inaccurate weapon, as was demonstrated in tests carried out by the Prussian army in 1782. At 100 paces 60 per cent of shots hit a 10 foot by 5 foot target but at 200 paces only 40 per cent and at 300 paces a mere 25 per cent.

A well trained infantryman could reasonably expect to hit an upright stationary figure at 80 yards, while 200 yards was the maximum range for engaging larger targets such as formed-up bodies of men.

0	50	100	200 yards

a Biting the cartridge open

b Priming the pan with a small amount of powder

c Pouring the main charge of powder down the barrel, followed by the ball and the wadding

d using the ramrod to pack down the charge

e pulling back the cock to full lock

f aiming and firing

The normal rate of fire for a musket was three to four rounds per minute, but to achieve even this under battlefield conditions was a considerable feat. The whole process of loading and firing was broken down into an extended series of movements, which were carried out together by the soldiers in a unit. The main actions are shown.

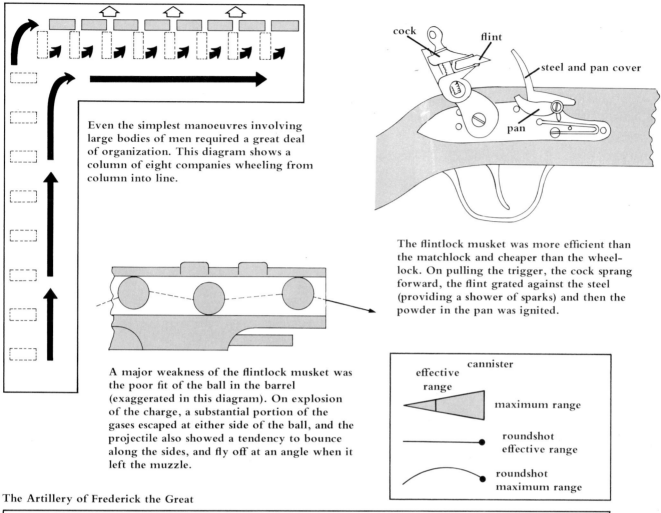

Even the simplest manoeuvres involving large bodies of men required a great deal of organization. This diagram shows a column of eight companies wheeling from column into line.

cock

flint

steel and pan cover

pan

The flintlock musket was more efficient than the matchlock and cheaper than the wheel-lock. On pulling the trigger, the cock sprang forward, the flint grated against the steel (providing a shower of sparks) and then the powder in the pan was ignited.

A major weakness of the flintlock musket was the poor fit of the ball in the barrel (exaggerated in this diagram). On explosion of the charge, a substantial portion of the gases escaped at either side of the ball, and the projectile also showed a tendency to bounce along the sides, and fly off at an angle when it left the muzzle.

cannister

effective range

maximum range

roundshot effective range

roundshot maximum range

The Artillery of Frederick the Great

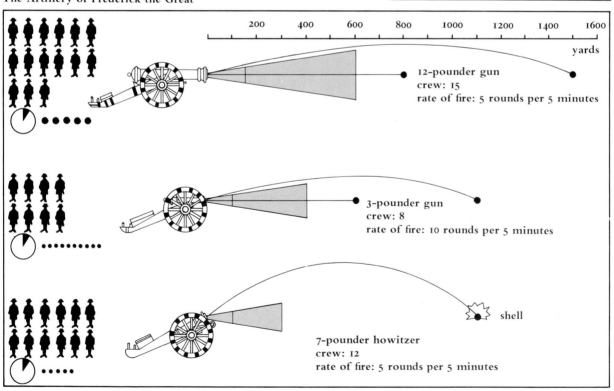

200   400   600   800   1000   1200   1400   1600

yards

12-pounder gun
crew: 15
rate of fire: 5 rounds per 5 minutes

3-pounder gun
crew: 8
rate of fire: 10 rounds per 5 minutes

shell

7-pounder howitzer
crew: 12
rate of fire: 5 rounds per 5 minutes

The brutal corporal punishments of the Prussian army were notorious. Here, one man runs (or rather walks) the gauntlet, while another is strung up and beaten.

battle; an improvement in musket design, the 'conical' touch-hole, was introduced to shorten loading time by forcing a trail of the powder charge into the pan when the charge was rammed home, making separate priming unnecessary. But whether the infantryman stood in ranks firing his musket volleys or charged the enemy lines with his bayonet, he faced a high risk of mutilation or death at the hands of a prepared enemy. He had, at most, two other soldiers between himself and the enemy, for the Prussian line was only three ranks deep; if one of the others was killed or wounded, he had to move up to preserve the front rank. That rank, of course, took the main brunt of fire from the opposing line, and sometimes case-shot from artillery pieces as well. But it was the soldier's duty to be shot at, and the situation was no better in other eighteenth-century armies. Frederick tried to lighten the risk for his native Prussians to a certain extent by throwing in the free battalions first; the latter were definitely expendable.

In view of the unenviable situation the recruit found himself in, his natural hope – and the officers' constant fear – was that he would be able to desert as soon as possible. Desertion was indeed a major problem for Frederick's army, as it was elsewhere. During the futile War of the Bavarian Succession in 1778–9, the army lost 3400 men in battle and more than 16,000 in desertions. Elaborate precautions were taken to prevent desertion, and no thought was given to consideration of the individual. Civilians were amply rewarded for apprehending deserters, and fined for failing to do so. Unreliable soldiers were accompanied by men of known loyalty while on sentry duty. For leaders of attempted mass break-outs there were severe punishments ranging from mutilations – by having ears or noses cut off – to execution by hanging or the firing squad, although punishment never seems to have made an appreciable difference to the level of desertion. One of the main duties of Frederick's cavalry while the army was on the march was to patrol the infantry's flanks, not so much to protect them against enemy attack but to prevent them from slipping away. For the same reason, the infantry were kept well away from woods and one of the main duties of officers was to accompany men when going to forage or to bathe; at times combating the enemy became secondary to combating desertion.

Harsh punishments were not reserved for deserters alone, however; they were the very essence of Prussian discipline. Frederick's view on the subject was simple: 'In general, the common soldier must fear his own officer more than the enemy.' This was seen at its clearest in battle, when the officers and NCOs patrolled behind the line to administer a short, sharp thrust with sword or half-pike at any soldier who turned his back on the enemy. 'If a soldier during an action looks about as if to flee,' Frederick wrote in 1745, 'the non-commissioned officer standing behind him will run him through with his bayonet and kill him on the spot.' However, the same treatment was meted out to those tempted to plunder the dead or wounded when the enemy was put to flight. Prussian soldiers were expected always to remain in line, even when victorious.

Discipline in camp was also harsh: officers beat or whipped men for slovenly drill on parade after the example of their royal master, while thieves were branded. The favourite punishment, however, was running the gauntlet. The standard number of whippers was one hundred, armed with soaked hazel wands. Lest the victim attempt to run down the tunnel and miss a few whippings, a sergeant walked backwards in front of him with a half-pike pointed at his chest. Meanwhile the corporals patrolled the outside of the tunnel to punish any whipper who did not lay on hard enough. And the miscreant was sometimes forced to run the gauntlet not once but twelve times, or even multiples of twelve on successive days. Most of those who suffered thirty-six runs over three days did not live to describe the experience.

So much for punishments; what of the soldier's rewards? Major-General Yorke thought the life of the ordinary Prussian soldier not to be too harsh. He considered their conditions 'better in that service than in any other, provided they can accustom themselves to the confinement of never stirring out of sight of their officers without an NCO with them'. Frederick took a close interest in his soldiers as well as his officers; promotion, like so much else in the Prussian army, was often at his personal whim. He kept an eye out at reviews for soldiers with ten or so years' experience and the proper bearing to serve as NCOs. These men received a half-pike, stick and gloves, and were treated with some politeness by the officers, who rarely beat them. The NCOs in their turn were polite to the officers and brutal to the men. From lance-corporal the NCO might advance to corporal, sergeant and sergeant-major, but there his progress stopped; there was obviously no place for men promoted from the rank-and-file in the Prussian officer corps.

Most of the time, the soldier's life revolved around a standard routine of guard duties, short (but very sharp) drill sessions, church parades on Sundays, and quite a high proportion of leisure time. Yorke wrote: 'Frederick never fatigues them unnecessarily, so that, when they have once learned their exercise, they have nothing to do but their ordinary duty.' For most of Frederick's reign soldiers were billeted on townspeople, although barracks were built later on. They were paid two thalers a month, which was hardly generous, and had to pay for food and for replacements to worn-out parts of the uniform. The original uniform of Prussian blue had, it was true, been issued free, but Frederick had reduced the cost of this item by cutting down the lapels, which meant that the coat could not be buttoned in cold weather. The king had also insisted on cheaper cloth, and done away with the cloak altogether. Soldiers were permitted to marry and live with their wives in billets, and wives might even accompany them on campaigns if they were prepared to look after the wounded and the laundry.

In view of harsh treatment meted out to the serving infantryman by Frederick, it might be assumed that after long years of service the faithful soldier could look forward to a quiet and secure retirement with a pension. Not so: if he was too old or infirm to stay in the army, the king rewarded him with a royal licence to beg (if he was Prussian; foreigners were merely expelled from Prussia with nothing at all). Not surprisingly, old soldiers tried to hang on in the ranks for as long as they could.

The two-year training period for the new recruit, and the yearly manoeuvres once he had returned to his civilian occupation, meant that the Prussian was a highly trained soldier. When this was combined with the resolute aggression and incisiveness of the commander-in-chief, the Prussian military machine could work wonders. The battle of Leuthen in 1757 during the Seven Years' War provides the best example of Frederick's genius at this time.

The Prussians were heavily outnumbered during this war by a combined force of Austrians, French and some of the smaller German states. Some of Frederick's units had taken a beating in the last days of November 1757, and the victorious Austrians, an army of 70,000 men, set up camp near Breslau. Frederick marched against them at the beginning of December, and he was evidently determined on an all-out gamble. He informed his generals:

> I shall attack, against all the rules of the art, the army of Prince Charles, nearly thrice as strong as our own, wherever I find it. I must take this step, or all is lost. We must beat the enemy or all perish before his batteries. So I think, and so will I act.

Then, reaching an unusual peak of emotionalism, he added:

> If any regiment of cavalry shall fail to crash straight into the enemy when ordered, I shall have it dismounted immediately after the battle and turned into a garrison regiment. If any infantry battalion so much as begins to waver, it will lose its colours and its swords, and I shall have the braid cut from its uniform.

Frederick himself was popular, in spite of the ferocious discipline he imposed on his men. His exhortation did, apparently, have great effect on his army, and the enthusiasm the troops felt for 'old Fritz' was rekindled; hardened veterans are said to have shed tears, although Frederick's main stimulus had been the stick rather than the carrot.

The king advanced his army toward the Austrian lines and began to deploy his left flank for battle against the enemy right. The Austrians saw the threat and reinforced their right with their infantry and cavalry reserve. But Frederick's move was a feint, and the moment he saw the Austrians respond he put the well-drilled manoeuvrability of the Prussian army to good use by executing a

massive right turn. This carried the bulk of his men, under cover of a ridge of high ground, straight across the enemy line to concentrate against their weakened and exposed left flank. The Austrian commanders were able to observe a part of this movement, and assumed that the outnumbered Prussians were shirking battle. 'The Prussians are off,' exclaimed an Austrian marshal. 'Don't disturb them.'

But even before the battle had begun the Austrians saw that they had been outmanoeuvred. Their cavalry made a spirited attack on the leading Prussian hussar squadrons, but a counter-attack with support from six infantry battalions annihilated the Austrian left wing after an hour and a half. A decisive element in this success was the Prussian artillery, still rather underestimated by Frederick. But he had developed highly mobile units based on the six-pounder gun drawn by teams of six horses, and these were able to move rapidly to support the infantry with devastating blasts of case-shot. He had

also had the heavy artillery from the fortress at Glogau mounted on wheels, and these, too, were used for infantry support.

Badly mauled by cavalry and artillery, the Austrians retreated into the village of Leuthen. Unfortunately, they made the mistake of concentrating their forces in the village, which only served to present a fuller target for the Prussian case-shot.

That evening at about five o'clock, when the sun was going down, the Austrians organized a cavalry attack against the Prussian left flank, but Frederick's cuirassiers and dragoons put them to flight. Darkness fell, and it was left to the Prussians to pursue the enemy as they fled in disorder. Total Austrian losses were some 10,000 killed and, in the battle and subsequent surrender, no less than 40,000 prisoners. Prussian losses were about 6000 dead. There is little doubt that Leuthen was the greatest of Frederick's victories. The cavalry had decided the final issue, but the infantry had been brilliantly organized and

The combination which led to the great Prussian victory at the battle of Leuthen. Frederick himself (left) was not only a great strategic and tactical innovator; he also had the ability to inspire his men, and to extract the best from his generals, even when, as during the Seven Years War, the weight of enemy strength seemed overwhelming. But his victories depended upon the fighting power of his troops: the steady discipline and marching power of the line infantry (above) and the dash of his cavalry (below). These forces commanded the admiration of all Europe.

had responded to Frederick's emotional appeal with verve, loyalty, and (most important) better-than-usual musketry. The artillery had also played a vital part, largely due to its ability to remain mobile.

In the years following the Seven Years War, the influence of Frederick and his army was far-reaching and profound. Throughout Europe politicians, military leaders and ordinary people watched with partisan enthusiasm as Frederick consistently took on and beat the overwhelmingly superior armies of his enemies. Each of his victories in turn was cheered in salons, inns and coffee houses, sometimes in countries whose own armies were being routed by the remarkable Prussians.

Inevitably the admiration for the Prussian armies led to imitation, and everywhere military thinking began to develop on the Prussian pattern. At its most trivial, parts of the Prussian uniform were copied; at its most profound, innovations like Prussian drill and man-oeuvres were adopted. Everywhere among the major armies of Europe – in France, Germany, Britain – the Prussian way was analyzed, approved, almost deified, often to the point of absurdity. The success of Frederick's army depended too much on his individual leadership; as Yorke wrote, 'the machine is created, subsists and is put in motion solely by the genius of the Prince that presides over it'. After Frederick's death in 1785 the Prussian army rested on its laurels, believing, as did its many admirers, that Frederick's own ideas could not be bettered. Frederick himself did not live to see the creation of a new style of army in which the ordinary soldier could advance himself through initiative and bravery and did not have to subordinate himself completely to the whim of his superiors. The king could not have countenanced this, and neither did his successors, until the battles of Jena and Auerstadt in 1806 destroyed the old Prussian army.

But the essence of the Prussian military system survived, at times below the surface, frequently reappearing and reasserting itself until modern times. What was the essence? It was not an expression of patriotism; the high proportion of foreign mercenaries in Frederick's armies rules out nationalistic fervour as a reason for their success. For some, the ethical standards of Lutheranism concerning the authority of the prince and the duty of the subject were influential, but that was only a small part of the story. It has more to do with the idea that the soldier must have no qualms about obedience, even when ordered to his death, and that the yardstick of constant obedience will keep an army (and society) together more surely than any fervour for a cause.

This same question had puzzled Frederick himself in 1740, when, a new and youthful monarch, he had reviewed the assembled army of 20,000 men before the invasion of Silesia. He remarked to his field-marshal, Prince Leopold of Anhalt Dessau, upon how strange it was that so many men, resentful, better armed and stronger than the king and his generals, should still shiver in their presence. It was, Leopold assured him, 'the result of order, discipline, and narrow supervision'.

11
FROM NATIONAL GUARD TO GRANDE ARMÉE

The French Revolution of 1789 brought about a fundamental change in the armies of Europe, a gradual transformation that began in France and spread outwards. The relatively small army of professional soldiers, many of them foreigners, who happened to be employed by the king of France in 1789 gradually became the massive force of men who were both conscripts and yet were still fired with a dedication to live and die for the Nation, the Revolution, and the Emperor – a truly national army both in composition and ideals. As always, other countries had to adapt their armies to keep up with the latest trends in warfare; even the great war machine built up by Frederick of Prussia could not compete. This time it was not new weapons or even new tactics which mattered, but a new motivation for the troops.

The transformation of the French army did not happen overnight, nor did it proceed according to a carefully thought-out and logical plan, any more than the Revolution itself did. Different ideas were tried at different times in response to new external or internal threats; some of them worked, and some of them did not. In order to see how order was brought out of chaos, we must examine the army's role in the Revolution and the Revolution's reorganizations of the army.

The royal army, in the early months of 1789, consisted of 102 regiments of foot and twelve extra battalions of light infantry, sixty-two regiments of horse, and seven of artillery. About a quarter of these units were foreign, mainly Swiss, German or Irish. Two-thirds of the officers were nobles, and the growing conflict between the nobles and the Third Estate in France was reflected in the army; there were several instances in 1788 when troops refused to obey orders, especially when called upon to deal with popular demonstrations against the royal government. Even the officers were not united, for the junior ranking officers resented the fact that they could never reach the rank of captain unless they were of noble blood. On the whole, however, the army functioned well enough; recruiting was adequate and desertions and dismissals rare – until July 1789.

A member of the Paris National Guard of the early 1790s, and (on the right) a soldier of the grenadier company of a Napoleonic infantry battalion.

On 8 July, in order to protect the city and the Revolution, the people of Paris asked the National Assembly for permission to raise their own guard of citizens, and on 13 July they decided to go ahead without waiting for the assembly's response. The soldiers of the royal army stationed in the capital, far from resenting the creation of this new and independent force, supported the idea, and several of them are known to have taken part in the storming of the Bastille on the following day in order to obtain arms for this Parisian militia. The regulars began to disappear from their ranks in order to join the militia, soon to be known as the National Guard; one regular unit, the Regiment of Provence (later the 4th Infantry Regiment) lost 121 men in July alone while stationed in Paris. During the autumn, as the political chaos ensured that pay fell into arrears and rations and supplies became scarce, men continued to desert the royal army, and when the Paris National Guard abducted the royal family from Versailles to the capital in October many of the noble officers felt their own positions to be threatened and began to leave the country. In 1790 recruitment was down by half while desertions and discharges continued; by the end of that year the army had lost 20,000 men.

As the regular army went into a steep decline, the National Guard was growing rapidly. In addition to the 48,000 men raised in Paris, similar units were recruited in other towns in 1789 and 1790. They were raised by, and composed of, the *bourgeois* for local defence against either royalist counter-revolutionaries, radical *sans-culottes*, or apolitical brigands. In June 1790 the organization of National Guard units was extended to every community in France, and it was specified that they should adopt the uniform of the Paris militia – red and blue for the city, and white for the king (the regular army infantry wore white only). Only 'active citizens' – that is, the wealthier ones – were supposed to be allowed to join, but in many cases the ranks were filled up with poorer artisans, workers and deserters from the royal army, who were happy to receive the better pay and softer discipline of the National Guard. The subalterns and NCOs of the regular army also saw their chance and were accepted as officers in the new force. But although the National Guard soon counted more than a million men in its ranks, they continued to denounce any suggestion that they should leave their cities and towns to serve in the general defence of the frontiers.

Throughout 1790 and 1791 the National Assembly continued to reform the regular army. In September 1790 all new second-lieutenants were forced to undergo a competitive examination – noble birth was no longer a sufficient qualification. A certain number of officers were to be chosen from among the NCOs, too. In January the traditional names of the regular army units were replaced with numbers in order to foster the idea of equality. Soldiers were allowed to join political clubs, and veterans with sixteen years' service were to be regarded as 'active citizens' regardless of how much they paid in taxes. In order to boost recruitment, a lump-sum bonus was authorized, to be paid immediately on entry. All of these measures encouraged the soldiers of the regular army to stay, and induced others to join, and the regular infantry increased by 14,000 in 1791. At the same time, noble officers were being given more and more encouragement to leave: in June 1791 they were forced to swear allegiance to the nation, the law, the king, and a constitution which had not yet even been published; in July all noble titles were abolished. By the end of the year about 4000 officers had left the army, and many of them had left France. They were replaced by NCOs or by volunteers from the National Guard.

The fall of the Bastille in July 1789 had been a watershed; the failure of the king's attempted flight from Paris in June 1791 was another. In addition to a great increase in the emigration of noble officers, this latter event led to the calling up of 100,000 National Guard volunteers for full-time duty until December 1792, for it seemed that the king had negotiated with foreign armies for an invasion of France to restore him to his rights. Now there were two armies in France, both charged with the defence of the nation against foreign invasion, and the two were quite different. The volunteers had responded to France's call for one campaign, the regulars for an indefinite period. The volunteers wanted to serve France and the Revolution, the regulars wanted good pay and the chance of promotion. The volunteers elected all their own officers, the regulars did not. And all this was in addition to the fact that the volunteers had better pay and easier discipline.

In the spring of 1792, Austria and Prussia finally declared war on France. The volunteers, though successful at first in the campaign in Belgium, were badly

The soldiers of the Revolution came from a variety of sources. The Parisian National Guard (below left) was a haven for deserters from the Royal army in 1789, and was also seen by the bourgeoisie as a defence against the lower classes. By 1792, however, the threat of invasion meant that radical sans-culottes (below) were volunteering in large numbers. By 1793–94, conscription was being enforced, and the men so recruited (right) were usually portrayed as brutal ruffians by enemies of the Republic.

trained, equipped and led, and they soon began to be pushed back towards Paris. The bad news from the north-east led to the invasion of the Tuileries and the deposition of the king in August, which in turn caused even more officers to emigrate; by the end of the year as many as half of the officers in the regular army were former enlisted men who had been promoted since the start of the Revolution. Another result of the defeats was a new wave of volunteers, including 40,000 from the border areas and 20,000 who marched out of Paris in September. This time there was no attempt to restrict recruits to the wealthier 'active citizens'; the country was in danger, and every man counted. At Valmy on 20 September the French army contained forty-four battalions of regulars and thirty of volunteers, fighting together – and they won, and went on to conquer Belgium that winter. But success brought its own problems. The rapid penetration of Belgium, Savoy and the Rhineland overstretched the supply lines just as winter was beginning, and at the same time the volunteers, with the campaign apparently won and their terms of service expired, went home.

Of course the war was not over, and on 1 February 1793 France's position became far more perilous when Great Britain declared war and organized a coalition of France's enemies, who soon included most of Europe. The government responded on 21 February by abolishing regiments (apparently even *numbered* regiments were thought to be élitist) and grouping the regulars and

volunteers together in 'demi-brigades' with a centre battalion of regulars and two flanking battalions of volunteers. Each battalion contained a grenadier company and eight companies of fusiliers. All now wore the blue coats of the National Guard. In theory, this amalgamation would give the reorganized French army the professionalism of the old regulars together with the patriotic enthusiasm of the volunteers, but the immediate reason for the change was the need for rationalization of supply systems and terms of service. Three days later the government called for another 300,000 men, and this time they were to be conscripted.

As it turned out, these two measures were not enough to deal with the dangers faced by the government in 1793. Aside from external threats, the introduction of conscription led to revolts within France. Finally, on 23 August, the government announced the *levée en masse*: 'From today until the time when the enemies have been driven from the territory of the Republic, every Frenchman is permanently requisitioned for the service of the armies.' In effect this meant the conscription of men between the ages of eighteen and twenty-five, together with a siege economy, fixed prices, an authoritarian government, and the Reign of Terror. Soon there were three-quarters of a million men under arms.

The huge, rather disorganized army of 1793–4, composed of various different elements – former regulars, volunteers and conscripts – rapidly dwindled in size. Inflation and inefficiency crippled all government

attempts to supply it properly, nor indeed, was the danger of invasion pressing after 1794. Spain and Prussia concluded peace treaties with France in 1795, and French armies overran the Low Countries. The rate of desertion mounted during the winters of 1794–5 and 1795–6, until there were probably less than 350,000 men under arms early in 1796.

This smaller army was, however, a very effective fighting force, as it was to show in the campaigns of 1796–7 in Germany and Italy. Its origins had given it a strength and solidarity which other European forces lacked – although it was still by no means a perfect fighting machine.

The first novel element in this French army was its political motivation. The armies in Italy and Holland, for example, expressed quite clear republican attitudes, and were well aware that they were very different from the armies to which they were opposed. The memoirs of various soldiers – Sergeant Fricassé or *Canonnier* Bricard for example – although perhaps describing men's motivation too benevolently, give a good idea of the atmosphere of the time. Fricassé relates how, although numbers were falling because of desertion during the harsh winters, they 'hardly regretted losing these dogs [the deserters], and thought of them as venom being extracted from a man who had been poisoned'. Bricard describes how the announcement of the *coup d'état* of Fructidor in 1797 (in which conservative members of the government were ejected, with the army taking a

prominent part) was greeted: 'this good news pleased the men, who had been watching with sadness as aristocracy seemed to be gaining over patriotism in France'.

Republicanism was not universal, however. Royalism was quite common, especially in the *Armée du Rhin*. A royalist agent wrote:

> We gave new pamphlets to the army, we gave out vast quantities of money, we even gave watches to the outposts . . . such was the spirit of the army that the soldiers trampled the tricolour cockade underfoot. The commonest phrase of the lips of the soldiers was 'the devil take the Republic'.

There were always problems whenever troops were shifted from the Rhine to Italy or vice versa, partly because the different political views in each army brought about clashes between the men.

The reasons for these differing political views may well lie mainly in the attitudes taken by the various commanders of the armies. For now that the armies were living in the poorer frontier regions of France or across the borders, a great feeling of solidarity had arisen. The conditions of 1795 had been so bad that the men who survived and did not desert really felt themselves to be a hard core of seasoned men. Their commanders, most of whom had risen to high rank since 1792, were often popular, having undergone many of the same privations (and found the same solutions) as the men: one divisional commander affectionately became a byword for pillaging – 'a robber like Erneuf' was the phrase. Young men like Desaix, Kléber, Hoche and Bonaparte were very popular. Indeed, Hoche's death in 1797 brought out many spontaneous demonstrations of grief.

The officers of this army were experienced and unified, too. When the great fall in numbers had taken place in 1795, many had had to lose their positions, and those that were left were generally of higher quality. Their backgrounds varied, but many had served as subaltern officers or NCOs in the old royal army; they knew the details of their trade thoroughly. They also felt more secure during 1796–7. As the familiar 'tu' was dropped they had to be addressed as 'Monsieur'; courts martial and the normal trappings of military discipline also began to return.

The soldiers were sometimes unhappy at the returning signs of a more conventional army, and some officers (Bricard mentions one who beat men with the flat of his sabre) certainly abused it. But the discipline was not of the niggling, minute Prussian kind. Punishment for minor crimes was not harsh; on the other hand those men

found guilty of major misdemeanours, such as abandoning guard duty or pillaging, were shot. In spite of any individual instances of friction, the soldiers still felt closer to their officers than to the civil populations of the lands they invaded; and by 1797, with most of provincial France in the grip of reaction and the army the symbol of revolutionary innovation and success, soldiers often found themselves unpopular in their home country as well.

The armies that invaded Germany and Italy in 1796 were, then, quite unified bodies, which had an *esprit de corps* based on their novel history and their sufferings over the previous two years. They were still ragged and poorly supplied, however, even when they reached the rich areas of Lombardy and the Danube valley. The government was notoriously lax about arrears of pay, for example. Even in 1798, the 33rd *demi-brigade*, posted to Mantua, had to revolt to obtain its pay. Having to live off the country-side rather than carry their own provisions, too, made French armies unpopular, and they frequently found themselves engaged in a brutal war against the local peasantry. In spite of attempts to stamp it out, one of Bonaparte's adjutant generals wrote of how, in April 1796, on the line of march in Italy, 'In all villages, hamlets and houses in the countryside, everything has been pillaged and looted . . .' Soldiers sometimes sold their weapons or shoes when the situation was very bad. Indeed, sometimes discipline threatened to break down altogether. Just before the battle of Arcola in 1796, for

Above *By the time of the battle of the Pyramids (1798) the French army was a tried and tested instrument; its ability to resist the furious charges of the Mamelukes by forming divisional squares was the key to victory.*
Left *French light infantry, a symbol of the new flexibility which the Revolutionary armies brought to warfare.*

example, a pamphlet called 'Disgust with our profession' (*Le dégout de métier*) circulated widely amongst Bonaparte's troops who were very badly provided for during the siege of Mantua. It poked fun at generals, and in spite of all efforts could not be suppressed. Then during the battle of Arcola itself, men refused to obey orders, and at one stage some soldiers deserted Bonaparte himself.

Yet these, badly behaved, starving soldiers fought superbly, partly because those elements which made them difficult to control – their political opinions, their lack of equipment, regular pay or provisions – made them clearly different from their opponents. As Baraguey d'Hilliers, one of Bonaparte's staff wrote: 'The soldiers live only for glory.' The phrase 'la gloire' constantly recurs in descriptions of the armies of 1796–7. Men who had remained with their units through the vicissitudes of the previous years were normally enthusiastic about the military virtues, and were prepared to take great risks if there was the possibility of promotion. When properly commanded and when their pay and provisions were assured (or when victory could

mean replenishing the war chest) Frenchmen were formidable. General Vaubois wrote in October 1796 that 'the French soldier will do anything if well led'.

The Prussian army and most other eighteenth century armies, had maintained rigid formations to force unwilling soldiers to stay on the battlefield; they had been unwilling to use troops in rough country or in loose formation for fear of desertion. In the French Revolutionary army, such considerations were much less important, and the army was much more flexible on the battlefield as a result. There has been considerable debate over the exact tactical dispositions of the revolutionary armies; any one proposal can be contradicted by others, and the reasons for success lie more in the versatility of the men than in formalized parade-ground evolutions. Wanting to fight, the soldiers found the best means of doing so whatever the conditions. Thus, the army certainly did use large groups of infantry as light troops, firing individually from cover; the Austrian forces in Germany and Italy, often firing in line from open ground found these men difficult to deal with. But the general disposition of the French infantry in *l'ordre mixte*, a combination of troops in three or four deep lines, and deeper, more manoeuvrable formations (generally a 'battalion column' about thirty metres wide and twenty or so metres deep) was also very important. The French army could be relied on to adapt itself much better than its opponents to difficult ground: and so in battles such as Arcola (fought in marshes) and Rivoli (in hilly terrain)

125

Austrian advantages in numbers or the late concentration of the French army could be compensated for. Eighteenth century warfare had developed into a stately minuet; the French Revolutionary armies, by introducing a new vigour and *élan*, cut through the previous conventions, and gave the men of the armies on the Rhine and in Italy a great advantage, whatever their shortcomings in other respects.

By 1805, the ragged, unconventional (and frequently undisciplined or mutinous) armies of the Republic had become the *Grande Armée* of the Emperor Napoleon, an army so effective in all respects that it broke through the restrictions of the previous European states system. French armies were to reach as far west as Lisbon, and as far east as Moscow; French generals were to be given kingdoms, from the southern tip of Italy to the north of Sweden. The foundation of this achievement, unsurpassed since the Roman legions, was the sheer fighting quality of the French soldier. Napoleon's achievement was to graft discipline, a strong loyalty to himself and to his empire, and a pride in unit or formation onto the enthusiasm and tactical innovation of the revolutionary armies to produce one of the best fighting forces Europe has ever seen.

The transformation of the revolutionary armies took place on two levels. The first began during the war of the Second Coalition (1798–1801). After the successes of the years 1796–7, France was attacked by another coalition in 1798, and her weakened armies, which had been steadily declining in numbers, had to be reinforced to meet this new threat. The 'Loi Jourdan' was the response to this need. Passed in 1798, it established the principle of regular conscription: 'Every Frenchman is a soldier and must defend his homeland.' All Frenchmen aged between eighteen and twenty-four were liable to serve, for three or five years. This law defined conscription during the Napoleonic period, and although it never brought in the number of recruits available (because of widespread evasion and desertion) it ensured that the Napoleonic army had a constant flow of new faces into its ranks, and was, by and large, all assembled in the same fashion, whereas, as we have seen, the armies of the 1790s had come into service by a variety of methods – either as old regulars, National Guards, volunteers (of various kinds) or members of mass levies.

This institutionalized mass conscription, then, provided the basis for Napoleon's armies. What brought these new recruits to the level of fighting ability shown by 1805 was the reorganization and training the army underwent from 1802 onwards. Victory over the Second Coalition was assured by 1801, and in the period following, the First Consul (soon to become Emperor) created the instrument he needed to give full flower to his military genius.

One of the major stages in this was the suppression of the political bias which had inspired the men but had often undermined discipline. The upper officers were under particular surveillance. Republican generals such as Richepanse and Decaen were sent to the colonies;

General Bernadotte's entourage was closely investigated, and some (such as General Donnadieu) were imprisoned. The secret police followed very actively a suspected conspiracy involving General Simon which was thought to be spreading left wing propaganda amongst the troops. Units from the royalist army of the Rhine whose fidelity was suspect were sent to San Domingo, and some regiments were given a suspiciously above average number of dangerous tasks. Many members of the *Association des Philadelphes*, who had never hidden their hostility to Napoleon's imperial ambitions, died at Wagram in 1809 when, under their leader, General Oudet, they were all in the same regiment which was annihilated by Austrian artillery fire.

Napoleon was not concerned with generalized political views so much as ones centring round particular individuals who might be a threat. The army as a whole thought of itself as a progressive force, rooting out the old triumvirate of kings, aristocracies and priests which ruled most of Europe, and although little was done to encourage this opinion, there was no attempt to turn the mass of the troops into a conservative, merely repressive force – although Napoleon might well have tried to achieve this had he thought he could succeed.

To replace any precise political affiliations, many of the formal trappings of military life were re-installed. As early as 1796, in Italy Napoleon had realized the importance of loyalty to a unit, had issued flags to the *demi-brigades*, and had given them mottoes to embroider on them ('I was happy – the 32nd was there' was a typical example). In 1803–4, regiments replaced the *demi-brigades*, uniform was standardized, and a complex system of awards for bravery was instituted. Elite units were once again brought into existence. The Imperial Guard was formed in 1804. Two thirds of its ranks were filled by veterans; it consisted initially of 5000 infantry and 2800 cavalry.

Jean-Baptiste Barrès joined the Imperial Guard in 1804. He was issued with underwear and footwear and his basic uniform: a blue coat, lined and piped with scarlet and still bearing buttons with the consular fasces, breeches and waistcoat of white jersey, a three-cornered hat with yellow cording, epaulettes of green woollen cloth and red tabs, and weapons and cartridge pouch. He was also ordered to grow his hair in order to make a pigtail. Finally, Barrès was instructed to sell the possessions that had not already been taken away from him – the army of Napoleon demanded all or nothing!

The Imperial Guard were proud of their uniform and determined to show it off to the best advantage. Jean-Roch Coignet resorted to the ploy of buying false calves in order to improve the shape of his legs, although he was contemptuous of Russian soldiers who padded their chests and shoulders with rags. Hair was worn according to strict rules: the queue was six inches long, tied with

A grenadier of the Imperial Guard, in full dress uniform. His weapon is the 1777 Charleville musket, an inaccurate weapon, but one which saw long service.

Napoleonic Tactics

The success of Napoleon's armies depended to a great extent on the flexibility and fighting power of the French infantry. By 1806, the French infantry had used the legacy of the drill books of the last years of the royal army, added to the revolutionary heritage of a new type of soldier, not dependent upon the stifling petty discipline of the *ancien régime*, to create a set of evolutions which enabled the troops to undertake most of the tasks they were called upon to perform. The basic formation for a battalion of nine companies was the battalion column – four companies deep and two across with one company detached to form a covering screen of light infantry across the front. In this formation the men could be moved swiftly without the need to maintain an unwieldy line. Before engaging enemy infantry, however, the battalion could easily deploy into line to present maximum firepower, all the time covered in this manoeuvre by the company of *tirailleurs* along their front.

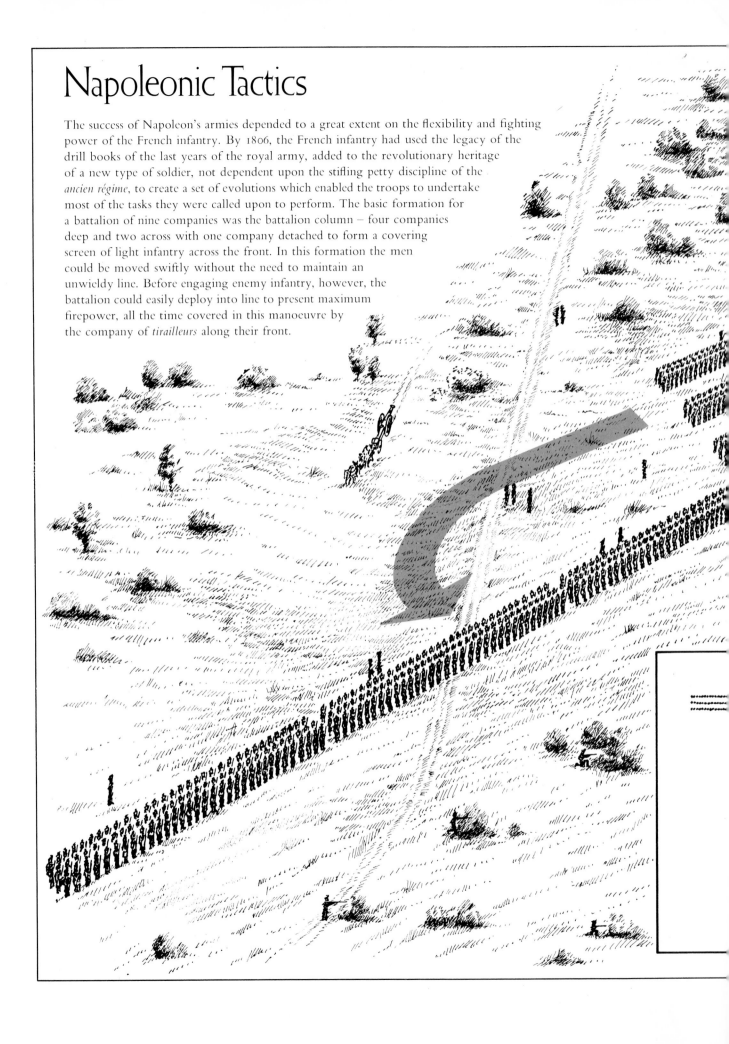

The square was the formation which gave the greatest protection against cavalry, and could be formed by simply folding back the line, as shown here.

A typical infantryman, with cooking pot on his back. The weapon is the 1777 'Charleville' musket, an inaccurate, unreliable flintlock.

black worsted ribbon with ends exactly two inches long. In addition the Guard wore powdered 'pigeon's wings' over each ear; every night the men put their hair in curling papers and in the morning a hairdresser came to the guardhouse to arrange their hair. The bearskin eventually replaced the three-cornered hat in the Imperial Guard. On parade it was a superb creation, but on the march it was a hindrance of the first order; it was designed to be carried in a sort of muff-box which, being made of cardboard, fell apart at the first good shower of rain.

Cavalry too, which had been the least effective part of the Revolutionary armies, now came into its own. The sartorial pretensions of the Napoleonic cavalry were legendary even in their own lifetime. The light cavalry inherited from the *ancien régime* their long waxed moustaches, which were curled and tended with extreme care, and now they were required to grow their hair in order to wear a queue and tresses. Great was the chagrin

of chasseur Charles Parquin when his youthful visage failed to produce a bristling moustache and his hair never seemed to grow! On the night before general inspection soldiers could not sleep for fear of spoiling their beautifully powdered coiffure. Finally in 1806 the men were ordered to cut their hair for ease of care on campaign – a sensible move which saved the men from vermin, although it caused a great many heartaches.

Napoleon's decrees of 1803 laid down the uniform for all the cavalry, and the regulations were altered very little throughout the period. Charles Parquin gives a graphic account of the uniform of the 20th Chasseurs in 1803. The shako was made of black cloth from which hung a 'flame' of golden-yellow cloth, ending in a point with a tassel; there was also a black and red plume attached to the front of the shako. The queue was four inches long, the penultimate inch being bound with black wool. Tresses were bound with small bands of lead and all the hair was pomaded and powdered. The green dolman had

Left Chasseurs of the Imperial Guard. These gorgeously caparisoned horsemen played a vital role in the French victory at Austerlitz in 1805.
Above One of the mannerisms for which Napoleon became famous. Such incidents helped to foster the Napoleonic legend in the ranks of the army.

golden facings and piping, white braid, and five rows of buttons, and the green breeches were also adorned with white braid. Sashes were eight inches wide, made of green and gold cloth with tassels. Chasseurs also wore cloaks slung from the right shoulder. Boots and gauntlets completed the uniform, together with a sabretache – a pouch attached to the sabre belt by three long straps to hang at the level of the left knee. It was used by chasseurs to carry despatches and, said Parquin, 'a handkerchief if they possessed one'.

So much for revolutionary simplicity. The *sans-culottes* had despised the breeches of the fastidious aristocracy, so concerned with correct form and good taste, and had mocked them for the long hours they spent on their *coiffeur* – and here were the men of the national army, missionaries sent abroad to spread the gospel of the Revolution, attired in costume more appropriate to the dance or the royal *levée*! They did not see it that way, of course. *Esprit de corps* and readiness to die for the nation were no longer the preserve of nobles; every man now had a stake in France and the Revolution and should be ready to fight gloriously and if necessary die gloriously.

In view of this, no self-respecting trooper would go into battle looking anything less than immaculate.

Loyalty to the unit and the wearing of new uniforms would have been of little effect, however, without the addition of a firm loyalty to Napoleon himself. It is clear that even as First Consul, Bonaparte had a cynical attitude towards his soldiers, and felt little real sympathy for them; but it is also clear that he knew how to appeal to them, to win their trust and affection. On parades and inspections he often claimed to recognize soldiers who had fought in earlier battles and campaigns. His mannerisms became famous – pinching huge grenadiers on the ear or sitting impassively with his head thrust into his chest during a battle.

In some respects, the Imperial armies were no better off than the Revolutionary forces. Accommodation was non-existent outside the *écoles militaires* and the barracks in garrison towns. The battalion of skirmishers of the Imperial Guard to which Barrès belonged was quartered at the *château* of Ecouen, and in 1802 Jean-Roch Coignet was quartered in the barracks of the Capuchins near the Place Vêndome. The Imperial Guard, to which both these men belonged, did best, but conditions were often cramped and uncomfortable. A trooper hardly ever had a bed to himself, even in hospital; Coignet's bed-fellow was six foot four inches tall and at least a foot too long for the bed; when Napoleon inspected the barracks early one morning, he had difficulty in believing that the head and the feet belonged to the same man. Having discovered that this was indeed the case, he ordered longer beds for the entire guard! These new beds, were over seven feet long, and cost more than a million francs altogether.

On campaign it was the task of the fourrier and brigadier-fourrier to secure billets for the men. They were issued with tickets and sent about two hours ahead of the main army to requisition accommodation, in accordance with the Revolutionary Army's custom of living off the land. It was cheaper but much more hazardous than organizing a huge supply train, but in Poland and Russia, where there was little shelter and no food to speak of, the policy came disastrously unstuck. Although it must be said that Napoleon did make provisions for supply trains and mobile bakeries, these proved inadequate to meet the needs of the growing French armies.

Food in barracks was usually adequate, depending on who was in charge of supplies. Our friend Coignet was at one point placed in charge of the NCOs' boarding house and given their mess contribution of 45.70 francs for food. He was forced to spend an extra 21.20 francs per day – 8.10 each for bread and wine, three for dishes furnished from outside, and one for wood.

The situation was quite different on campaign. Rations were often scarce – sometimes non-existent – and the men became expert at marauding. In Poland all food was concealed. Sometimes hams, potatoes, rice and flour were discovered in caches left by the absent inhabitants of villages where the French rested; after

Eylau twenty-five wagon-loads of provisions were found buried in the forest and gardens. These included salted meat, flour, rice, potatoes, honey and preserves, as well as tools and linen. At other times the French were less lucky; during the retreat from Moscow in 1812 they had to drink their horses' blood, while in Spain they lived on rice, roots and thistles. In friendly country and good weather they might eat well, driving herds of animals along with them on the march, while at other times they were reduced to eating horses, dogs, rats and rotten cabbages.

Pay too, was a problem; in 1806 it was five months in arrears. While on campaign, it was of little importance to the soldier – he had nowhere to spend it, and he found that he lost heavily if he converted from one currency to another. For what it was worth, however, Jean-Baptiste Barrès was paid twenty-three sous a day in 1804. Of this sum, nine sous were deducted for his mess bill and four sent to a fund to provide underwear and shoes; the remaining ten were issued every ten days as pocket money. Corporals received thirty-three sous a day and sergeants forty-three. In all cases, replacement uniforms had to be paid for individually. Occasionally there were bonuses: after the birth of the king of Rome a review was held and a gratuity of a litre of wine and twenty-five sous was distributed to those who took part.

Although better dressed and usually more regularly paid than his republican predecessor the Napoleonic soldier used basically the same drill book (that of 1791) and the same evolutions. At the beginning of the Empire, in 1804, drill was good. The soldiers, particularly those in the Imperial Guard, took great pride in their smart drill and worked hard at it. Jean-Roch Coignet, small of stature and stentorian of voice, was a drill sergeant who, like others of his rank, was ordered by Napoleon to drill alone in order to perfect the movements before putting the other ranks through them. Later, each NCO had to take his turn issuing orders to 100 men. This was well and good while in garrison, but in the field there was no time to learn the art of wheeling in a straight line as for a parade in Paris; one learned as far as possible the use of one's weapons and trusted to Divine Providence and good feet for the rest. And in any case, the huge losses in the catastrophic campaign of 1812 led to a high proportion of raw conscripts in the army the following year, causing a decline in the precision of drill, as well as other problems: the surgeon-general found that recruits were wounding themselves and each other through lack of sufficient weapons training.

The infantryman's weapon, the flintlock musket, was the 'Charleville' model which had been in service since 1777. A trained soldier could fire about two shots per minute on the battlefield; a raw recruit rather less. Because of this and because of the risk of malfunctions, it was essential that each musketeer should also be a 'pikeman', able to defend himself with the bayonet. The range of the musket was not great, either; being a smoothbore weapon it could only be accurate up to about 55m (sixty yards). At a distance of 125m (150 yards) it might score a hit if fired by an experienced man at a large target such as an advancing column.

Discipline varied in harshness according to the commanding officer. Less free and easy than during the Revolution, it was, nevertheless, less onerous than in most contemporary armies. Absence at roll-call meant a spell in the guard-house; fines were also common. Courts martial were formal affairs, with several cases being heard at one sitting. An officer sitting on the court had the right to defend an accused man if he had no one else to speak for him, and this gave the man a good chance of obtaining a fair hearing. Often, however, justice was meted out in a more arbitrary form: soldiers caught *in flagrante delicto* were sometimes shot. With provisioning so poor, there were frequent small scale mutinies, although nothing on the scale of some of the demonstrations during the Revolution.

In spite of the changes, Napoleon's army retained the same *élan* and fighting spirit of its predecessor. Partly, no doubt, this was because of the possibility of promotion and bravery on the battlefield was one of the main criteria for advancement. But it was also partly because of the direct connection with the revolutionary tradition. In 1805, a quarter of the *Grande Armée* had fought in the Revolutionary Wars, and a further quarter in the campaigns of 1800. Almost all the officers and NCOs had had some combat experience in these wars.

The combination of hopes of promotion, a strong tradition of bravery and offensive action, a tested and flexible set of evolutions, an experienced officer corps and a trusted commander made the French army a formidable opponent in 1805. Perhaps the best illustration of the power of the army lies in the activities of Davout's corps in the battle of Austerlitz in 1805 and the battle of Auerstädt in 1806, when in performing two very different tasks, they acquitted themselves superbly.

At Austerlitz, Davout's corps had two main tasks. The first was to get to the battlefield on time; they were held 100km (seventy miles) away near Vienna in order to lure the Allies into attacking a supposedly weak French army. The men were forced-marched to the battlefield when it was clear that the Allies had taken the bait. III Corps began to arrive on the evening of 1 December 1805, and those that could fell into an exhausted sleep. Their role in the battle the next day was a defensive one. They had to hold the French right flank against the main Allied attack until Napoleon's counter-attack in the centre brought victory. It was a dark, misty morning when at 7 a.m., the first shots were fired. A desperate sanguinary struggle developed around the villages of Telnitz and Sokolnitz. There was little opportunity for manoeuvre and Davout's exhausted men were facing a determined attack by fresh troops. The villages were taken; a counter-attack failed, and in the confusion, some French units fired at each other. But in spite of their tiredness, and in spite of fighting hundreds of miles from their homeland against heavy odds and (seemingly) about to be cut off, III Corps continued to hold out. More elements of the corps were arriving; at 10 a.m., the

An engraving of the battle of Austerlitz. In the foreground Rapp announces the capture of Russian standards, while in the background Davout's men drive the enemy into a lake.

French counter-attacked more successfully, pushing through Sokolnitz and then, when pushed back in their turn, continuing house to house fighting in the village. By the late morning, Davout had stabilized the position, despite the odds, and by now the French attack through the allied centre had ensured victory. But Davout's men had to continue fighting until 5 p.m., when the Austro–Russian forces were finally crushed.

At the battle of Auerstadt fought on 14 October 1806, Davout's men showed a different set of skills. While Napoleon and the bulk of the French army was defeating a smaller Prussian force at Jena, Davout's men, 27,000 strong, placed themselves in the path of the main Prussian army, 50,000 strong. From 8 a.m. to 12.30 p.m., Davout held off a series of Prussian attacks, by using the tactical expertise of his men. They deployed into line and resisted the Prussian infantry in a fire fight; they formed squares to hold off Prussian cavalry charges, and changed the angle of their front to resist outflanking manoeuvres. They then took the offensive themselves, and forced the disorganized and demoralized Prussians to retire. Prussian attacks on Davout's corps had not been well co-ordinated, but with their overwhelming advantage in

numbers they should have carried the day easily, and would have done so against troops less tactically adept and confident in their commanders than the French. As it was, III Corps suffered twenty-five per cent casualties.

At Austerlitz, then, the men of III Corps showed that they had the basic physical properties of soldiering – to march until they almost dropped and to fight on doggedly in appalling conditions and against great odds, while at Auerstadt, they demonstrated the technical skills of changing formation to suit circumstance and keeping in good order in spite of very heavy casualties where lapses would have been fatal.

The fighting man of Napoleon's armies of 1805–6, then, was certainly one of the best ever. He had inherited the best characteristics of the Revolutionary soldier, and with a mastery of all the arts of his trade had shown that, especially at Auerstadt, his *élan*, bravery and republican spirit were a far better basis for an army than the grinding discipline which was the Prussian legacy.

12
BILLY YANK AND JOHNNY REB

The troops who fought in the American Civil War were a distinct breed of fighting man, and unlike any the world had seen before. Firstly they were often enthusiastic volunteers whose expectations of glorious feats of arms were confounded by the technical innovations which changed the face of the battlefield, making it an altogether grimmer place than they could have ever imagined when they joined up. Then, the troops had been educated in a democratic egalitarian society and the habits of deference and obedience which traditional military discipline had demanded were alien to their outlook. And although they were fighting in a civil war, they were, in general, prepared to pursue the struggle without any of the bitterness and atrocities which so often render such conflicts more horrible than international war.

The direct preparation for war began late in 1861, when the Confederate States of America came into being. By March, 35,000 men had responded to the Confederate call to arms. In order to suppress the secession of the Southern States, President Lincoln had little choice but to call for volunteers to supplement the tiny federal army, 16,367 strong and severely affected by the decision of some of its best officers, such as Robert E. Lee, to join the Confederates.

Inspired by dreams of adventure and military glory, and roused to fever pitch by pleas to serve the nation in its hour of need, recruits on both sides made a spectacular dash to arms. The response to Lincoln's call to arms in the North was overwhelming. So many regiments were organized that they could not all be equipped, and therefore some had to be turned away. A medical examination was soon introduced to weed out the unfit — and the women, a number of whom managed to join up in the early days of the war without being detected.

The militia was called up by presidential decree to serve for three months, but volunteers in the North were asked to sign on for three years. The splendidly attired

men of the pre-war military clubs were not, however, incorporated into the volunteers of the Union army; most of their societies were designated as militia units and soon lost their identities. Nevertheless, there were differences in uniform between the forces raised by the various northern states, some of whom were even dressed in grey. The officers of the regular army were not spread out among the new units. Instead of having a cadre of experienced leaders throughout the army, the militia and volunteers had to rely on either the officers of the former volunteer societies or the occasional retired regular. This accounts, at least in part, for the relatively poor performance of the Union army early in the war.

In the South, on the other hand, all the soldiers were

Left A Union infantryman and a Confederate cavalry trooper, in the blue and grey uniforms of the two sides. Right These smartly uniformed soldiers, posing against a painted background, typify the illusions about the war which characterised the attitudes of most early recruits.

volunteers, and the regular officers who had resigned their United States commissions were distributed among the raw recruits to leaven their inexperience and provide the vitally needed cohesion within the new units. Between the various regiments, however, there was not cohesion but rivalry. The volunteer units on both sides were raised on a state-by-state basis, and the men were accustomed to rally around their own state flag. When, later in the war, the remains of regiments were amalgamated, soldiers took it ill that they should suddenly have to mess and fight with men they regarded virtually as foreigners, and morale dropped sharply. In the same vein, troops disliked moving too far from their home territory. The men from the west of the Mississippi, particularly Texans, resented having to cross the river. It was the same problem – soldiers locally raised and retaining local loyalties – that the Parliamentarians faced in England in 1645 when they founded the New Model Army on a national basis to fight a national war. In the Confederate army, volunteers who signed on for twelve months were obliged to equip themselves, but those with three-year contracts were kitted out at government expense. A few were so committed to the cause that they equipped themselves anyway and refused to take any pay.

In these early days of war, volunteers units on both sides adopted nicknames to express their identity and (hopefully) their valour: 'Oxford Bears,' 'Union Clinchers' and 'Detroit Invincibles,' or 'Louisiana Tigers' and 'Racoon Roughs'. In the South, these names might be emblazoned on the men's caps. Ethnic units might display the emblems or dress of the old country, as seen in the shamrock-bedecked flags of Colonel Corcoran's 69th New York or the feathered *bersaglieri* hats of the 39th New York ('the Garibaldi Guard'). But both sides had their colourful Zouave-style infantrymen and no actual French connection was required for membership of picturesque units like the 7th Louisiana or the New York City battalion of firemen with their red caps, short jackets, sashes and baggy trousers, although ability to grow a moustache and goatee beard was an advantage.

There were many differences in culture, way of life, economy, and political principles between the North and the South, but the enthusiasm and expectations of the recruits on both sides were much the same. They knew what war was like, for they had read all about the exploits of Napoleon and of the British in India; indeed many of them had served in volunteer companies for several years, wearing splendid uniforms similar to those of the élite European forces. They had paraded, drilled, sung military songs, attended banquets, inspiring their townsfolk with pride and admiration. It was felt that once the enemy saw them, they might not even dare to start fighting – and that even if they did, one battle ought to settle the issue. How could the rebels hope to withstand the attack of the splendid 'Zouaves' of the New York Fire Department? How could the Yankees hope for success when faced with Virginian gentlemen, who thought of themselves as the heirs of the Cavaliers?

All civil wars are supposed to be short, of course, and most armies follow the military fashions of the successful fighting machines of their own day. But there are several aspects of the American Civil War which stand out, showing just how 'American' it was. First, as we have mentioned, the citizens of the United States loved to form themselves into volunteer companies to parade and drill in peace-time, in pursuance of the clause in the Bill of Rights guaranteeing the right to keep and bear arms. Even when these organizations were formally incorporated into the state militia, they still elected their own officers in miniature versions of congressional campaigns, complete with hustings, speeches, rounds of drinks and election promises. Then, too, most of the soldiers who fought in the Civil War had had the benefit of an elementary education, and they have left a great quantity of diaries and letters describing their experiences. Although by no means all of them boasted the same peculiar ideas about grammar, spelling and punctuation as this Confederate soldier, his letter home after the battle of Shiloh is fairly representative of the way soldiers expressed themselves on paper:

April 17 the 17 run Yank or die Yankey paper Yankey paper Corinth Miss April 17.

Dear Wife: i take the opitunity to rite you a few lines to let you know i am well at this time and i hope these few lines may come to hand and find you enjoying the same blesson . . . we was in a battle on the 6 and 7 day of April we got one cild two woned the Captin was woned slitly in the arm i come out saft i was . . . sceard at first i tell you it was not . . . [what it] was cract up to be i was in the tenth miss regt i saw plenty of yankeys out ther i thout every minet was the las the ball whisle around me worse than ever bees was when they swarm but I am saft yet this is some them paper i got in there camps i got a par shoes i . . . [got an] overcoat it is caintuckey geans and a larg nife five pound two foot in the blaid well i could a got eneything i wanted if i coulda toated it.

The volunteers and new recruits to the Federal and Confederate armies may have been enthusiastic to begin with, but their ardour soon cooled. Army life did not suit many Americans; the unwonted discipline, harsh living conditions and shocking casualty toll were all hated. But the great majority of soldiers, although realizing the horrors of war, became efficient fighters. European observers tended to despise the American armies because they did not conform to the norms of discipline and regulation dress found in more traditional forces; but lack of attention to these externals did not prevent the soldiers from holding firm on the battlefield. By 1865, American armies were at least the equal of any European forces.

The virtue of independence, enshrined in the political principles of every American, was highly prized by both 'Billy Yank' and 'Johnny Reb'. It is understandable, then, that the regimentation of life in camp, the necessity of learning drill, and the fact that there were masters who had to be obeyed came as a shock. There were about a

136

Confederate soldiers – hardly in standardised uniform but well equipped compared with their fellows in the later stages of the war, when replacement clothing ran out.

dozen bugle calls a day for the infantry but nearly twenty for the artillery and cavalry in the Union army, and the routine was very similar in the Confederate army.

For all arms, the day began at 5 a.m. in the summer and about an hour later in the winter. The first call was for the assembly of buglers, which also served to wake the men. They rose at once and dressed, since if they were not on parade when the assembly of soldiers was sounded fifteen minutes later, they would find themselves doing extra duty or a spell in the guardhouse. Once on parade,

reveille was sounded and the roll-call took place.

For the cavalry and artillery, the stable call was the time for grooming and feeding the horses. Half an hour after reveille came the breakfast call, known as 'Peas on a Trencher'. At 7.15 a.m., when sick call was sounded, all those requiring medical attention lined up outside the regimental surgeon's tent. (Whatever his ailment, the Federal soldier would probably receive the universal panacea of a dose of quinine.) The next call, for cavalry and artillery only, was the time for watering the horses. This was not necessarily the simple job that it might appear, especially when water was foul or scarce; even though the camp might cover several square miles, there were very large numbers of animals to be cared for. After

137

the battle of Antietam, McClellan's army had 38,800 horses and mules to feed and water.

Fatigue call came next for all arms. Quarters had to be tidied, ditches dug, wood cut, horse shelters built, and the carcasses of horses and mules buried – for just as disease killed more men in the Civil War than the enemy did, so it was with the animals. Assembly of the guard usually followed at 8 a.m., when the company sergeants appointed the guards for the next twenty-four hours, paraded and inspected them, and sent them to their posts. In some regiments, however, guard mounting took place in the evening after the beating of retreat. The guard mounting was an occasion of some ceremony, with music being played during the assembly, parade and inspection. The normal period of duty for a guard was two hours in every six, but they were exempt from fatigues or roll-calls while on duty.

The next bugle call indicated the time for drill to begin. For the infantry this lasted until 12 noon, although the cavalry and artillery had a supplementary call known as 'Boots and Saddles' for the mobilization of men, beasts and hardware on the manoeuvre area. The dinner call, 'Roast Beef', was followed by a certain amount of free time, more drill, and preparation for the dress parade, as well as another watering and stable call. At 5.45 p.m. the call for attention was sounded, followed by the assembly and retreat. The roll was again taken, and in the infantry there was a dress parade and inspection which was open to visitors, especially high-ranking local officials. Notices were read, drill was performed, and the companies were inspected and dismissed. Then came supper, followed at about 8.30 p.m. by attention, assembly and a final roll-call known as 'Tattoo'. At 9 p.m. 'Taps' was beaten on the drum, after which 'all lights must go out, all noises cease, and every enlisted man be inside his quarters'.

Union army regulations dictated that a camp should be laid out in a prescribed fashion, by companies, with the tents in files facing a street. This system was rarely used by volunteers, although the regulars adhered to it. Confederate camps were less well organized; for one thing, there were never enough tents to accommodate the army, and many men spent at least some of the war either bivouacking in the open or constructing their own form of shelter.

There were several different types of tent common to both sides: the Sibley or bell tent, which held twelve men; the wedge tent, which held four men comfortably and six if necessary; and the tiny dog tent, which was really only a length of canvas 1.6m by 1.4m (5 foot 2 inches by 4 foot 8 inches) with two holes for poles (not provided) and with buttons and buttonholes on three sides. This last provision enabled several men to attach their tents together to form a more complete shelter. Officers had a small wall tent to themselves, while large tents of the same design were used as field hospitals, housing between six and twenty patients each. In cold weather it was possible to heat a tent by digging a trough in the floor, lighting a slow-burning fire in it, and then covering the trough with stones; this was known as California heating.

In addition to tents, two other sorts of construction were widely used in the field. One was the bomb-proof or dug-out, most common along siege lines or within artillery range of the front. These were low buildings made of wood and earth, backing onto some structure such as a fortified wall or a rock-face. More common, however, were the cabins made by soldiers when they went into winter quarters and in which they spent the cold season in more or less complete inactivity. The descriptions of the shelters made on both sides testify to the extraordinary fecundity of ideas when the men were confronted with a non-military concept like building and furnishing a hut. The names given to these shelters – 'Astor House', 'Willard's Hotel', 'Five Points' – were indicative of the high regard their builders had for them, rather than the quality of life enjoyed therein.

On the Confederate side, where the shortage of canvas was more acute, the men built log huts in the style of pioneer cabins, with a stone fireplace as the vital centrepiece. Any number of men, normally between four and ten, might combine to construct such a dwelling. It was also possible to adapt tents (where available) by putting a fireplace at one end. Another method was to dig a hole under the tent and live partly underground. Finally, the Confederates used natural features such as ravines, covered over and floored in to form 'gopher holes'.

Union soldiers also built log huts, but more often used their tents as roofs for them. There were usually four men in each hut, since it took four dog tents to form a roof. A central fireplace was surrounded with walls of logs and mud, which made a mortar of varying efficiency depending on the skill of the maker.

The furniture inside these huts depended on the ingenuity and resourcefulness of the occupants. Ration barrels and boxes could be used in many ways, the former serving as chairs or chimney tops, the latter as tables or chairs. Bunks were constructed of ·logs or boards, and occasionally barrel staves were used for the bottom, these being pliant and therefore more akin to a mattress than the hard ground on which the men slept in summer.

Candles were the main source of light, but candlesticks were by no means uniform in appearance. One of the most popular was a bayonet stuck into the floor or table, with the candle secured in the socket. When the supply of candles faltered, the men used 'slush lamps' instead: these simple devices were merely sardine tins filled with grease from the cookhouse, with a piece of rag dipped in one end to serve as a wick. The lamp was then suspended from the roof or a wall bracket with wire from a hay bale. Although the grease was probably stale and smelly, this type of lamp provided plenty of light.

When it came to food, the Union soldiers were better off than their southern counterparts. They also received more than the British, the French, the Prussians, the Austrians or the Russians, but that did not stop them complaining:

> The soldier's fare is very rough,
> The bread is hard, the beef is tough,
> If they can stand it, it will be
> Through love of God, a mystery.

Of course, there were times when supplies were plentiful and times when they were scarce, but the theoretical daily ration consisted of twelve ounces (340gm) of pork or bacon or 1 pound 4 ounces (567gm) of salt or fresh beef, plus 1 pound 6 ounces (627gm) of soft bread or flour or 1 pound (454gm) of hard bread or 1 pound 4 ounces (567gm) of cornmeal. Officers were not fed on the same

rations as the men; they were given a cash allowance instead, and bought provisions for themselves and their servants direct from the brigade commissary. Field officers also had a fodder allowance for their horses.

Veterans of the Civil War had two culinary memories that caused them to wax eloquent, the hard bread (hardtack) and the coffee. Hardtack was the usual bread ration, and consisted of crackers made of flour and water which were also known as 'teeth dullers' or 'sheet-iron crackers'. Although sometimes eaten straight, they were often ground down and re-made into something more palatable. The memories of coffee were more favourable, and its mildly stimulant character made it an important part of the soldier's diet. Any halt on the march would provide an opportunity to brew, and as soon as the order came every fence-post in the vicinity would disappear into the fire. The coffee ration was distributed with the sugar and the men usually kept the two mixed together, but milk was a rare luxury.

Left *A group of Union soldiers outside their tent. The variety of uniform worn in the field is clearly visible. Smartness counted for little once the fighting began.*
Above *Three Confederate soldiers captured at the battle of Gettysburg in 1863. This defeat marked Lee's last chance to attack deep into northern territory.*

The decline of Confederate fortunes was reflected in the soldiers' rations. By January 1863, Lee's troops received only 18 ounces (510gm) of flour, 4 ounces (118gm) of bacon, and sporadic supplies of rice, sugar and molasses each day. The splitting of the eastern and western states of the Confederacy after the fall of Vicksburg in July 1863 meant that supplies of food could not be exchanged between the regions, and Confederate troops came to rely on beef and cornbread. Coffee in particular became very scarce. During sieges, of course, the situation became still worse: at Vicksburg and Port Hudson the men were reduced to one biscuit and two mouthfuls of bacon a day, and they were eventually forced to eat mule meat.

On both sides, rations were supplemented with food parcels from home. These might contain anything edible – even poultry and fruit pies – although sweets, vegetables, pickles and preserves were more common. Items of a general nature were also sent, as we see in this list of goods sent to John D. Billings by his family: 'Round-headed nails (for boots), hatchet (to cut kindling, tent-poles, etc.), pudding, turkey, pickles, onions,

pepper, paper, envelopes, stockings, potatoes, chocolate, condensed milk, sugar, broma, butter, sauce, preservative (for boots).' Shirts, as well as stockings, were often sent. But if, as sometimes happened, camp was struck while the long-awaited parcel was *en route,* the chances were that parcel and addressee would never meet.

Aside from foraging – discouraged by both sides, although it was difficult to prevent in the cavalry – the only other regular source of supply was the sutler. His wares ranged from sewing kits (for those who had the inclination to repair their uniforms) to meat pies (which, though generally reviled, were eaten in large quantities). The sutler was usually the most unpopular man in the army, partly because he was a civilian, but chiefly because of his prices; these, although not beyond the pockets of most soldiers, were generally considered exorbitant. Occasionally, tempers would flare and a sutler would be 'cleaned out' – that is, a party of soldiers would plunder his stores and make it impossible for him to continue trading. The authorities often turned a blind eye to such incidents. Local farmers and pedlars also traded directly with the troops, but they did not provide serious competition to the sutler with his government contract.

It is generally believed that the men on the Union side wore blue while the Confederates wore grey. The regulations of the two armies certainly stipulated these colours, but the actual turnout was somewhat different. On both sides the demand for uniforms far outstripped the supply, particularly in the early days, and the recruit was often clothed by his own community. The enthusiasm of local ladies for doing their part towards the war effort often outdid their skill, and many of the uniforms they produced were of non-regulation hue, ill-fitting and of poor quality.

As with most logistical questions, the problem of uniforms was greater on the Confederate side. It seems ironical that the cotton-growing areas should find themselves deprived of cloth, but the problem was partly of the Confederates' own making: government policy dictated that a large proportion of uniforms should be kept back for use 'in an emergency'. At one stage of the war, while Lee's men went without vital new issues of clothing, 92,000 uniforms were being kept in store. Some Southerners wore captured Federal uniform, although this could be disastrous in battle; this practice was especially prevalent in the cavalry. General Nathan Bedford Forrest eventually ordered that all captured uniform must be dyed.

Even worse than the cloth shortage in the Confederate army was the lack of footwear. The men were often forced to go barefoot through lack of leather. Some tried to make moccasins out of cowhide, but these had no strength in the sole and soon fell apart.

Militia regiments, especially the ones based on peace-time volunteer units, continued to take pride in their own colourful uniforms, but these were designed for the parade ground rather than the battlefield and soon wore

142

The heroic, romantic view of the fighting in the Civil War was not dispelled by the experiences of the early battles. The myth of heroic assaults died hard, and was a constant feature of artists' impressions (left). The battle of Missionary Ridge in November 1863 (above) was a vain attempt to storm Confederate positions in Tennessee.

rubberized layer which effectively protected the wearer from the weather, even on the march.

The soldier could function without greatcoat, knapsack or even shoes, but without a weapon he was useless. Important changes in weapons were taking place during the Civil War: first, the smooth-bore musket was being superseded by guns with rifled bores firing the Minié ball, and secondly the later stages of the war saw the introduction of breech-loading repeating rifles in the Federal ranks. These novel weapons caused much consternation among the Confederate soldiers, since there was no longer a pause for reloading along the Federal line. A Union soldier remarked that 'they say we are not fair, that we have guns that we load up on Sunday and shoot all the rest of the week'. In fact, however, the greatest number of bullets that a repeater could take at one time was sixteen and seven or eight were more common.

The vast majority of ordnance supplies were held in states loyal to the Union; early in the war there were only 150,000 weapons in the Confederacy, many of them out-of-date guns converted from flintlock to percussion. The Confederates had to set up completely new industries or buy from abroad, both of which they did, but at first many units were equipped with shotguns, fowling pieces, or even guns marked 'Tower of London' and left over from the war of 1812. The most rewarding source of re-equipment was the Federal army; weapons captured after major battles such as the Seven Days, Harper's Ferry and Fredericksburg, enabled the Confederates to build up a surplus by 1863.

By 1862, the weapons on both sides had become predominantly the same: the .577 Enfield rifled musket, of English origin, and the .58 Springfield. The latter, though of wider bore, was more popular because it was lighter. The similarity of calibre meant that the same ammunition could be used for both weapons, although the Enfield could become clogged with over-use of Springfield ammunition. Both guns were still muzzle-loaders, with an awkward and time-consuming loading process involving biting the end off the cartridge and ramming the ball and charge, both of which were potentially dangerous operations if the soldier was inexperienced or nervous. As with all muzzle-loaders, no doubt a certain number of ramrods were inadvertently fired in the heat of action. Another danger lay in failing to bite off the end of the cartridge so that the powder could escape into the barrel; the usual reaction to a misfire caused by this was to ram the charge down still harder, thus causing a genuine jam. If the weapon was not then discarded, the only means of rendering it usable again was to use a searcher to withdraw the jam and then to clean the barrel with boiling water and oil – not a pleasant task.

The cavalry on both sides used short rifles. The Confederates relied on the Enfield short rifle, difficult to deal with on horseback even though the ramrod was permanently attached to the barrel by a swivel mechanism. Rivalling the Enfield in popularity was the muzzle-

out. Afterwards they, too, were forced to rely on the poor-quality government issue, when available.

Recruits normally arrived in camp loaded down with all sorts of impedimenta lovingly packed by family and friends, but the soldiers of both sides soon learned to select only the most basic items to carry on the march. As soon as spring came and the men moved out of winter quarters, a vast amount of clothing was simply discarded. On both sides one of the first items to go was the knapsack, along with much it contained. The remainder, including such things as eating irons, cup and plate, and a change of shirt, socks and underwear, were wrapped in the blanket, carried across the shoulder and secured at the hip. This, together with the haversack of rations and his weapons, constituted the load of a veteran on the march. Of course, it left the soldier unprepared for steamy, week-long rainstorms, and lack of spare clothing meant that even in midsummer the Confederate light infantrymen were never really dry. 'Dry' became a euphemism for 'damp', while 'wet' meant 'drenched'. Union troops, however, had the advantage of one important piece of clothing not issued to the Confederates: a poncho with a

The American Civil War

The American Civil War was a time of transition in warfare. For the average infantryman, the battlefield suddenly grew larger, and entrenchments such as those shown here became an essential part of actions in the field because of the technical developments shown below. Efficient rifling gave most soldiers a vastly improved weapon, and breech loading began to take over for artillery.

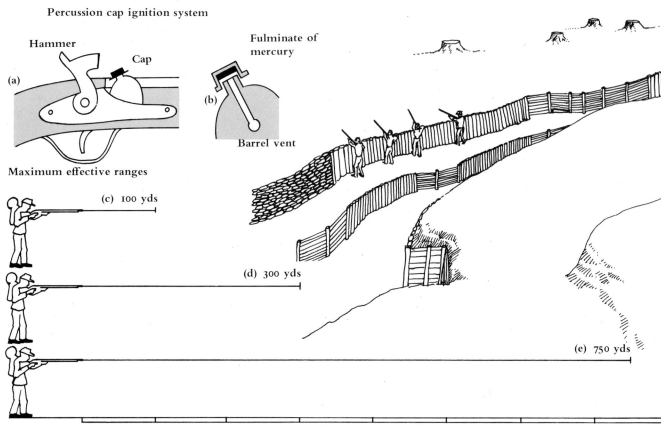

Percussion cap ignition system

Hammer

Cap

(a)

Fulminate of mercury

(b)

Barrel vent

Maximum effective ranges

(c) 100 yds

(d) 300 yds

(e) 750 yds

Musket balls and bullets

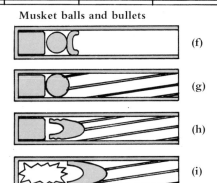

(f)

(g)

(h)

(i)

During the early 19th century most armies changed from the flintlock to the percussion system of ignition, which greatly reduced the number of misfires that were such a feature of the flintlock. Once the trigger was pressed the hammer (a) struck the copper cap crushing the fulminate of mercury and sending a flash down the hollow nipple to the barrel vent to ignite the main charge (b). The range of the infantryman's weapon was greatly increased through the widespread introduction of rifling and improvements in bullet design. The smoothbore musket of the early-19th century had an effective range of only about 100 yards (c). Rifled weapons did exist, but although their range was greater they were difficult to load and fouled easily (d). The 'rifle musket' of the American Civil War was a reliable and accurate weapon, by contrast, with a range of 750 yards (e). The improvement in the long range accuracy of the rifle musket was largely due to developments in bullet design. A smoothbore musket ball (f) had a loose fit and was held in place with a wad. The rifled musket ball of the early 19th century (g) had to be rammed down in order to grip the grooves of the rifling. This made loading slow and the rifling deteriorated rapidly. But in the Minié principle (h) the conical bullet was dropped loosely into the barrel. When the charge exploded the bullet expanded, gripping the rifling as it was propelled up the barrel, and ensuring a good gas-tight fit. Thus, proper ballistic stability and ease of loading were combined.

Trajectory of a low velocity bullet

50 yds

Danger Area 75 yds

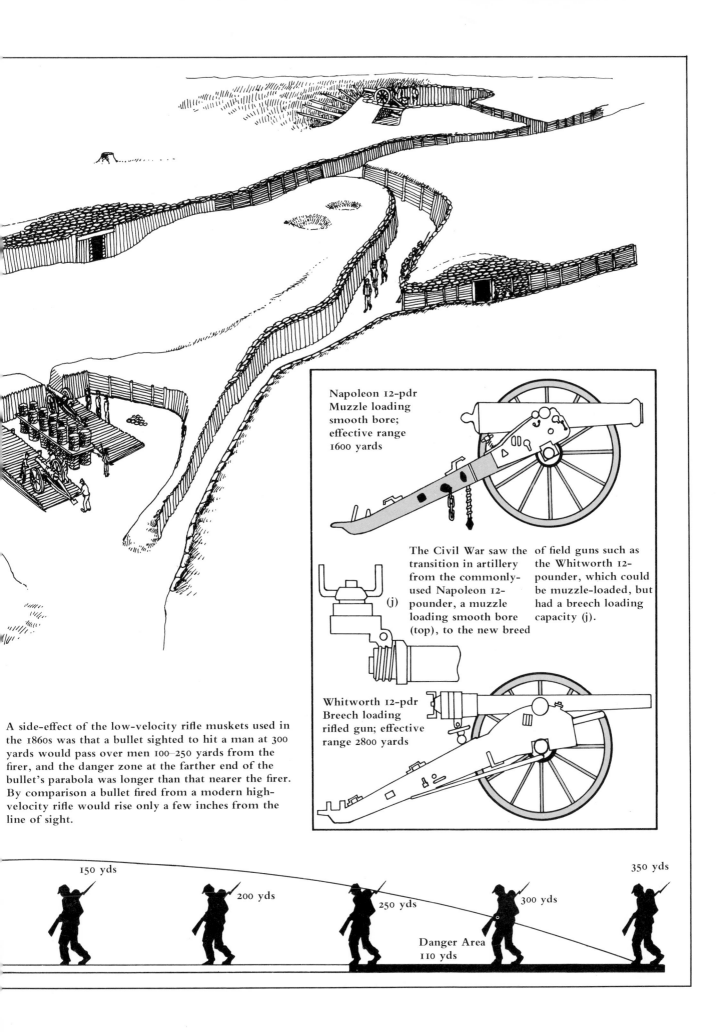

Napoleon 12-pdr
Muzzle loading
smooth bore;
effective range
1600 yards

The Civil War saw the transition in artillery from the commonly-used Napoleon 12-pounder, a muzzle loading smooth bore (top), to the new breed of field guns such as the Whitworth 12-pounder, which could be muzzle-loaded, but had a breech loading capacity (j).

(j)

Whitworth 12-pdr
Breech loading
rifled gun; effective
range 2800 yards

A side-effect of the low-velocity rifle muskets used in the 1860s was that a bullet sighted to hit a man at 300 yards would pass over men 100–250 yards from the firer, and the danger zone at the farther end of the bullet's parabola was longer than that nearer the firer. By comparison a bullet fired from a modern high-velocity rifle would rise only a few inches from the line of sight.

150 yds

200 yds

250 yds

300 yds

350 yds

Danger Area
110 yds

loading double-barrelled shotgun, used extensively by the Confederate cavalry even though it fired only buckshot. In the second half of the war, the Federal cavalry were issued with seven-shot Spencer and sixteen-shot Henry repeating carbines. One infantryman in Sherman's division also paid an extra $35 for a Henry rifle in 1864. A certain number of these weapons were captured by Southerners, but they had difficulty in obtaining the right sort of ammunition. Also in wide use were revolvers, especially the .44 and .36 Colt 'six-shooters'.

Bayonets were issued to both infantry and cavalry, the former using mainly the angular type while the latter received the sword type. There seems to have been a general aversion to this weapon, particularly on the Confederate side, since it got in the way of the loading process. In any case, on the rare occasions when the two sides fought hand-to-hand, the men tended to use their guns as clubs.

Pikes, bowie knives, swords and sabres were also used on both sides – the pikes were introduced into Lee's army in 1862 during a severe shortage of guns, but success in capturing Federal weapons made their continuance unnecessary. Bowie knives, more generally carried on the Confederate side than the Federal, were also intended for use against the enemy, but were probably more often employed for culinary purposes. They were issued to Federals, but not by the government; the town of Ashby, Massachusetts, voted early in the war that 'each volunteer shall be provided with a revolver, a bowie knife, a Bible and shall also receive ten dollars in money'.

Financial rewards from the governments were neither particularly generous nor prompt, although here too the Federals were marginally better off. At the start of the war a private received $13 a month, rising to $16 after 1 May 1864. The Confederates received $11, raised in June 1864 to $18. These figures, however, are deceptive. On both sides, pay was commonly in arrears – often by as much as six months – and prices were continually rising. The Confederates were harder hit since their currency was constantly being devalued, usually being practically worthless by the time it arrived, and they were often obliged to sell belongings to obtain enough money to live. Union soldiers, although they were also paid late, at least received currency that was worth something; the monthly pay muster, which took place on a Sunday, must have been one of the few occasions when soldiers actually looked forward to being reviewed and inspected. They also had one further advantage, which helped alleviate the worries of men with families: the paymaster could keep back a specified amount of the soldier's pay for forwarding directly to his family. This process, known as allotment, bypassed the need to use the unreliable postal system.

Having received their hard-earned pay, the soldiers would while away the long hours in camp by gambling. Cards and dice were the most popular forms; the men's propensity to gamble was deplored by the high commands of both sides, but would simply go underground

if banned. A more visible problem was alcohol; more than half of all courts-martial during the war on the Confederate side were due to drunkenness. If liquor was banned in camp – and it usually was – the men would smuggle it in or patronize the nearest saloon. Occasionally, in cold or wet weather, whiskey was officially issued to the men, but only in derisory quantities. The only official source of whiskey in any quantity was the medical depot, for it was the favourite anaesthetic and disinfectant. The medical officers, of course, had easy access to it, and this occasionally aroused resentment in the rest of the camp. Pedlars also sold liquor in camp, often of doubtful origin, but the men drank it nonetheless, even if a jigger cost a dollar.

Women, too, were available to relieve the men of their cash; the number of camp followers greatly diminished as the war went on, but feminine company was usually on hand in towns near the camp, for a price. But not all the soldiers' amusements cost money. Both sides boasted a number of regimental bands, while a banjo, fife or fiddle might accompany the men in an impromptu singing session around the campfire. Reading and writing were also popular pastimes in these literate armies; the troops eagerly seized upon newspapers, religious tracts, novels, or any other reading matter that came their way, and they wrote innumerable letters home. Hunting, baseball and mock battles with snowballs helped relieve the monotony of life in winter quarters.

In the early days of the war, there was a prevailing attitude of leniency and lapses of discipline on both sides. It was not, in the long run, conducive to the maintenance of either good order or morale, but it was understandable. Officers, many of them new appointees with no military training or experience, were not accustomed to command their social inferiors as their European counterparts were, and the newly enrolled soldier had not been brought up to defer to anyone's opinions. In some of the peace-time volunteer units now serving as militia, the officers had been elected by the men; whether this made the men more or less ready to obey them is debatable. Certainly it is true that insubordination was one of the most common offences of Civil War soldiers. Then there was the question of lawbreaking outside the camp; in the early days, such cases were supposed to be tried by the civilian courts, and it was only in October 1862 that the Confederate authorities, faced with the complete disruption of the civil court system, began to deal with such cases.

As eventually standardized, the discipline and the forms of punishment – still lenient by European standards of the time – were similar on both sides. The Confederates had two types of court-martial: the special type, composed of three officers, which tried non-capital offences committed by privates, NCOs and subalterns,

Above right *Union soldiers resting after drill.*
Right *Confederate guerrilla fighters fire on Union vessels using the Mississippi river.*

and the general court-martial, which had greater jurisdiction and was composed of between five and thirteen officers, all of whom outranked the man on trial. When delays and dismissals of cases in the civilian courts became intolerable, because of the inability to find witnesses, in October 1862 the special military court was set up; this consisted of a colonel, a judge-advocate and a captain.

The Federal system was slightly more sophisticated, possibly because the machinery of military discipline already existed in the small regular army. The regimental or garrison court-martial could fine the offender a month's pay or sentence him to a month's hard labour or prison. The general court-martial dealt with capital offences within the camp, as in the Confederate army, and the military commission tried men for crimes committed outside. Appeal from these bodies lay with the judge-advocate-general or the president, and in 1862 Congress passed a law requiring all cases resulting in prison sentences or execution to be referred to the president; Lincoln took a personal interest in several cases. The law slowed down the execution of sentences to such an extent that the provision was later modified to allow army and departmental commanders to review certain types of cases.

By far the most common offences were those that could be punished at company level, either by the captain or a sergeant. Sleeping at one's post, desertion, cowardice and assault on an officer, had to be referred to higher authority, but offences punishable at company level included absence without leave, disrespect to senior officers, petty theft, absence from roll-call, disturbance after the beating of Taps, sitting while on guard, gambling, drunkenness and leaving the sentry beat without relief (known as 'taking French leave'). Punish-

A wounded member of one of the Zouave regiments on the Confederate side. Medical services on both sides were poor, and deaths in hospital were common.

ments might be extra guard duty or latrine digging, or various forms of public humiliation: riding the wooden horse, wearing a ball and chain, with a weight varying from 12–18 pounds (5.4–8.1kg), marching around camp carrying a log, standing on a barrel for a certain number of hours while carrying a knapsack full of stones, carrying a bag of sand, and so on. Confinement with a bread-and-water diet was another common punishment. More brutal expedients such as whipping, branding or head-shaving were far less common. Executions were rare, but when they occurred they were treated with ceremony, and the whole camp was turned out to witness the malefactor's end.

Bad living conditions (especially for the Confederates) and an onerous drill and discipline were not, however, the worst problems the soldier had to face. The infantry-man of both armies was the victim of a cruel fate. Technology, in the shape of the rifled musket, had outstripped tactics, and the finely-dressed lines, squares and columns of Napoleonic warfare were now out-moded; men could be shot down at ranges of 275–365m (300–400 yards) instead of being difficult to hit at over 90m (100 yards). Infantry now had to be entrenched to fight with the maximum effect, and although this was realized, officers trained to expect a mobile battlefield found it hard to cope with, especially if they were taking the offensive. As a result, casualties were very high and the front ranks suffered terribly. Public opinion was shocked when photographs of dead troops at Antietam in 1862 were published, but the soldier had to face such sights every time he went into action. Out of a total force of 85,000 men, Lee's army suffered 20,000 casualties in 1862 at the Seven Days battle; later that year, at Second Bull Run, a Federal army of 73,000 men suffered 14,400 casualties, and the Confederate army of 55,000 lost 9400. During the battle of Gettysburg, when desperate Con-federate attacks plunged forward, one Federal unit, the 1st Minnesota, lost eighty-five per cent of its strength, and the Confederate 1st Texas regiment suffered eighty-two per cent losses. Of the total Confederate army of 75,000 men at Gettysburg, 20,451 were casualties.

That the fighting man of the American Civil War was prepared to endure battles in which such losses were incurred speaks volumes for his ability. Both sides contained units of outstanding character, and even when badly led were prepared to undertake seemingly hopeless tasks.

As for the wounded, they had to rely on a medical service which started at zero and improved rapidly during the war. It is yet another indication of the amazingly sanguine hopes of Northerners in the early days of the war that the Federal army at the first battle of Bull Run had neither hospital tents nor surgeons. When the battle did not lead to the swift collapse of the rebellion, however, medical services were introduced, with field hospitals, ambulances and a separate ambul-ance corps for specialists. The Confederates failed to adopt such a sophisticated system, choosing instead to detail certain fighting men – usually the least fit – to

Confederate dead at Fredericksburg in May 1863. Such slaughter was a result of the technological improvement in the weaponry available.

remove the wounded to the rear. Many, of course, were never rescued and died on the field from exposure or suffocation, while others were recovered and treated by the enemy. In hospital, the wounded man might find himself surrounded by victims of smallpox, dysentery, measles, malaria and pneumonia – diseases that claimed the lives of four times as many soldiers as the enemy. Bad diet, hygiene and weather took their toll.

More features united Billy Yank and Johnny Reb than divided them. They spoke the same language, chewed the same tobacco, sang (mostly) the same songs and played the same games. The 'Rebel yell' became famous in battle, but Yankees yelled too. The Confederates recruited Red Indians to their flag; so did the Federals. Whenever the two sides were camped near each other the soldiers fraternized; groups of men would meet in no man's land to exchange goods – coffee for the Con-federates, for whom it became a luxury, and tobacco for the Yankees. Sometimes they visited the trenches of the opposing side and played cards or dice.

The high casualties during the Civil War were not the result of a strong feeling of malice towards the enemy, but were exacerbated by the changing methods of warfare. Soldiers of the Union and the Confederacy were both Americans, and they had much in common.

13
POILUS AND
OLD CONTEMPTIBLES

Less than a week after the German invasion of Belgium on 4 August 1914, the 160,000 men of the British Expeditionary Force under Sir John French crossed the English Channel and took their place on the left flank of the five French field armies near the town of Mons. These two armies were the products of totally different military traditions. The British force was small, professional and well trained – probably the finest body of men ever to leave British shores; the French force was large, conscripted, and on the whole poorly trained. In France the concept of the 'nation at arms' had held sway since the days of the French Revolution; in England the idea of conscription was bitterly resisted until as late in the war as February 1916. The Kaiser is reputed to have referred to General French's 'contemptible little army'. In the eyes of the French generals, the BEF was no less contemptible, for by comparison with their own armies it was a drop in the ocean.

Both of these armies, the one confident of its professional standards, and the other convinced that its *élan* would carry the day, were inspired by a patriotic fervour, and both fought bravely in their different ways. But neither was equipped for new forms of warfare, and within a year both had changed radically. The BEF was too small, a precision instrument ineffective in the hammering of trench warfare, while the French army found that its traditions of relentless offensive, and of warfare fought in an exciting blaze of glory which covered all other deficiencies were no match for the technology of shrapnel and the machine gun.

Britain had traditionally relied on her navy for national defence and shied away from committing her army to European entanglements. The BEF of August 1914 comprised six infantry divisions and one cavalry division. Each of these infantry divisions deployed 12,000 men in twelve battalions, plus four thousand gunners equipped with four 60 pounders, 54 eighteen

pounders, and eighteen 4.5 howitzers; the remaining 2000 men included mounted troops, engineers, signallers, a supply train and transport train and field ambulances giving a total of 18,000 all ranks. Each battalion was equipped with only two machine guns. A British cavalry division comprised four brigades, each of three regiments, plus artillery, engineers, signals and medical unit – a total of 9000 all ranks.

The French army was founded on a series of conscription laws passed from 1872 onwards. From the age of twenty-one a young Frenchman was required to serve three years in the ranks, usually in the infantry. General de Gaulle has captured the atmosphere of French military obligation well:

Left A British Tommy and a French poilu; the men on whose shoulders rested the weight of the allied war effort in the enormous new conflict.

Right The regulation dress of the French army in 1916, a smartness which was not reflected in the clothing worn in action in the trenches.

True to his race, he (the conscript) appreciates the picturesque, unforeseen, exciting side of army life; he sings on the march, is keen on the range, throws out his chest during the march past – charges with deafening shouts. Nevertheless he counts the days, believing in his innocence that his discharge means liberty.

Any conscript who thought this was in for a shock: after his first three years he passed into the army reserve for another seven years and finally into the territorial army for six years. His obligations were still not over because he then moved into the territorial army reserve for a

Left *Frenchmen returning from the line in 1914, weighed down by their personal equipment and showing the strain of the First World War battlefield.*
Below *The raw material of the pre-1914 professional army of Great Britain. Volunteers assemble outside a recruiting office in one of the poorer areas of London.*

further nine years. In all, a Frenchman was liable for military service for something like twenty-five years of his life.

The most important administrative unit in the French Army was the corps of 44,000 men; the corps staff co-ordinated the complicated mobilization process. Corps comprised two divisions each of two brigades, and these in turn were formed of two regiments. The conscripts were mobilized by region: each man had his orders telling him where to report on the date of mobilization. All he had to do was walk to the nearest railway station, board one of the 4278 trains that had been allotted for this complex operation and report to his regimental HQ. Before 1914, the French General Staff predicted that something like thirteen per cent of reservists would refuse to report for mobilization. Actually, less than one-and-a-half per cent did so. Enthusiasm for war was so infectious that 3000 deserters from the peacetime army returned to their regiments to take part in the fighting. A

French mother was heard to declare in the opening days of the war: 'Not for anything in the world would I see my two sons out of it.' She was soon to change her mind.

Whereas British soldiers were recruited from the least privileged sectors of British society (many were illiterate, and thought of the Army as a haven from poverty) the French Army represented a more even social cross-section. French soldiers were mainly of peasant stock, hardy, strong and tough. They were big men. The British tended to think of the French as small, but in fact the average French soldier was usually stronger than his British counterpart, probably a stone heavier and two inches or so taller.

British soldiers were volunteers and joined the army out of choice, not because they were told to. Why was this? They signed on for seven years, partly because life in the Army was perhaps preferable to life outside. Men born in the slum tenements of Lancashire or the lowlands of Scotland were familiar with the hardships of military life – rats, lice, cold, lack of clothing. In the army they escaped from their worries and were told what to do; they did not have to think for themselves. Neither did they have to worry about where the next mouthful of food was coming from. During the winter of 1914–15, the men in the trenches were issued with over a pound of bread or biscuit per day, the same amount of meat (frozen, tinned or preserved), plus vegetables, cheese, pea soup, butter, milk, jam, tea, sugar, tobacco, and rum. At the turn of the century, statistics collected by Rowntree in York showed that the rations supplied to a soldier each week were greater than the average weekly purchase of a labouring family of eight. It is a striking reflection on the quality of British rations, that France, the nation of *haute cuisine*, neglected the feeding of her soldiers. Cooking was primitive and the food of poor quality; hot food was a rarity and usually sold by private contractors at extortionate prices.

Another reason for joining the British army was the intense feeling of *esprit de corps* and comradeship offered by the regimental system. A soldier's regiment (or battalion) became his home and the focus of loyalty and affection. He was taught regimental traditions and made responsible for maintaining its record of valour and duty. His social life was restricted to drinking or playing cards with fellow members of his battalion. British soldiers fought for 'King and Country' but the honour of the regiment came a close second. Accordingly, a high degree of 'peer group' discipline existed in the old, professional British Army. During the First Battle of Ypres a man in the Royal Welch Regiment kept his head down during a German assault and fired his rifle into the air; he was threatened with death by his comrades if he refused to stand up and take proper aim. By not fulfilling his assigned role in the operation of his platoon he was jeopardizing the lives of his fellow soldiers and betraying the honour of the regiment.

In France this tradition was completely absent. The fighting reputation of a French regiment was far more important than its historic traditions. Some regiments

did claim a historical lineage – mainly from the old royal regiments of the *ancien régime*, and two actually claimed as their ancestors two regiments of Irishmen who had fought for Louis XIV. The most important unit in the French Army was not the regiment but the company of 200 men composed of four sections. This was the backbone of French fighting power because it was considered to provide the optimum conditions under which the intimate relations between French officers and their men could be cultivated. The company was commanded by a captain and three lieutenants. These looked upon their men with an almost paternal affection and often referred to them as *mes enfants* – sentiments that would have been regarded as preposterous in the British Army.

British officers were interested in their men but their emotions were more restrained. Each battalion of the Army was commanded by a lieutenant-colonel, assisted by a major, an adjutant, a quartermaster, a regimental sergeant-major and clerks. The adjutant was the most important single officer in the battalion because he was responsible for the day-to-day running of the unit; he wrote out the orders and ensured that the junior officers and other ranks were properly turned out. Within each of the four companies of the battalion, certain men were assigned special duties, such as service in the officers' mess or with the transport; the fighting strength of the company was thereby reduced to about 200 and was subdivided in platoons of fifty men and sections of ten. When the mobilization order was given on 3 August 1914, the men who had served seven years with the colours were recalled by the Regular Army Reserve, in which they had been placed for five years after finishing active service. The reservists were filtered back into the battalion structure of their old regiments – they were not used to create new units. The Territorial Army was also called up. (In the French Army the professional officers looked down upon the reserve officers in the same way as in Britain the territorials were looked down upon.)

In British regiments the junior officers were usually unmarried and were expected to give their life to the regiment. Their life centred around the officers' mess where they were served good food and drink and indulged in conversation of a rather mediocre quality, usually about sport, but never about 'shop' (military affairs) or women. Officers were expected, nevertheless, to take their obligations seriously and concern themselves with the honour of the regiment and the health of their men. Officers were 'gentlemen' because they were brought up to command; the men, in the same way, expected to be commanded. It was felt that officers promoted from the ranks would not command the same respect from the men, an assumption proved utterly wrong in the holocaust to come. Certainly, officers had to have a private income as their social demands were quite beyond the resources of their pay, especially in the cavalry regiments. A second lieutenant received just under £100 per year; in the infantry he would need double this and in the cavalry ten times the amount. Occasionally, a long-serving soldier would secure a commission. Examples are more numerous in the artillery which was less fashionable than the other arms. John Hoggart, a gunner who had enlisted in 1893, won a commission in November 1914 and eventually reached the rank of major. Similarly, Battery Sergeant-Major D.S. Kempton was commissioned on the field after the Battle of Mons and eventually rose to the rank of captain; though mentioned in despatches and recommended for the Military Cross this was refused. Prejudice against 'rankers' tended to wilt, however, once the great expansion of the British Army got under way in 1915.

In France, officers had similar privileges to their counterparts in Britain. In garrison towns they were much sought after for social functions, great favourites with the ladies, and trusted by local tradesmen; a young lieutenant was considered a desirable son-in-law, a man of honour with 'prospects'. Officers in the British and French armies were also united in their distrust and contempt for 'politicians'; quips and jokes about them were legion. Yet despite this, relations between French officers and their men were not rigidly stratified as they were in the British Army. Pre-war policy had demanded that candidates for St-Cyr (the French equivalent of Sandhurst) should spend at least one year in the ranks. Thus junior officers knew their men well, and their relations with them were based on first-hand knowledge and not just good intentions. It was traditional to address soldiers with the personal '*tu*'. Further, this relationship survived the casualties of the first battles because large numbers of educated Frenchmen, who would have received immediate commissions in the British (or German) Army, chose to serve in the ranks, and in 1915 these were promoted. One such man who went to war in 1914 as a chauffeur was rebuked thus: 'Your grandfather commanded armies, you drive generals'.

The qualities demanded of a good soldier in the British Army included smartness (personal hygiene was taught to some of the recruits for the first time), good marksmanship, steadiness on parade and on the battlefield, and endurance in marching. Individual training was received in the winter months, training by companies followed in March, battalion exercises during the summer and army manoeuvres in the autumn. Tactics were supposedly taught, though too frequently neglected. They were based on fire and movement: part of the unit would charge forward under covering fire from those behind to prepare a new firing line about 180m (200 yards) from the enemy; they would then charge home with the bayonet in extended order.

the attack. Dashing forward shouting '*Vive la France*', '*En avant!*', or '*A la baïonnette!*', the excited soldiers were supposed to strike such terror into the hearts of the enemy that they would drive him panic stricken from the battlefield. French soldiers were deployed in mass formations with no attempt at concealment. Bayonets were fixed; flags fluttered in the breeze; regimental bands struck up *La Marseillaise*; officers wearing white gloves drew their swords and ordered the charge at the sound of the bugles. As an earlier French general had remarked of British tactics in the Crimea: '*c'est magnifique, mais ce n'est pas la guerre*'. These idiotic tactics were suicidal and the disastrous losses that the French Army suffered in the first months of the war were due to a complete misappreciation of the effects of modern firepower.

Failure to obey orders was punished by court martial or by the administration of summary justice by commanding officers in both the French and British Armies. Methods of doing this differed, however. Discipline in the French Army was stern but it was not overbearing and did not seek to humiliate the offender, as was the practice in the British Army. For cowardice in face of the enemy, desertion and mutiny, the punishment was death in both armies. In the British, lesser crimes of a serious nature were punished by No 1 Field Punishment – the dirtiest and most unpleasant tasks followed by two hours per day chained to a wagon wheel for up to twenty-one days – a practice that was utterly abhorrent to the traditions of the French Army. British officers were not vindictive, but men charged were usually convicted. One battalion that fought in the Battle of the Aisne in September 1914 was punished by its CO for a lacklustre performance by one execution for cowardice and No 1 Field Punishment for all stragglers. Although relations between French officers and men were certainly more paternal, a more draconian atmosphere set in under the strain of the failures of 1914–15. On 6 September 1914 the French Government passed a decree setting up *cours martiales* to try on the spot men accused of looting, desertion and self-mutilation. The courts consisted of seven officers who could convene the court immediately and punish offenders once they were found guilty. Clearly, this system was open to abuse and put conscripted men at the mercy of officers with the rank to convene a *cour martiale*.

An example of the dangers of this system were revealed in what came to be called 'the affair of the four corporals of Suippes'. In March 1915 the 21st Company of the 336th Regiment refused to join an attack at Perthes-les-Hurlus in Champagne. The divisional commander was furious and ordered the commander of the 21st Company to send four corporals and sixteen privates on a suicidal mission. Under the stream of fire that met them these men took cover in shell holes and returned to their own lines after dark. The divisional commander

Whatever the weaknesses of training in the British Army it was incomparably superior to that in the French Army. When in the reserve men were subject to two one-month periods of training during the whole of the seven year period; men in the territorial army received one period of two weeks' training, and the ageing veterans of the territorial reserve got only one day in nine years. This was quite inadequate, and in August 1914, General Lanzerac complained bitterly that 'our men are not sufficiently trained at present to stand up on equal terms to the Germans'. This paucity of training in part explains the simplistic French tactics of 1914. These were based on what was called the *offensive a l'outrance*, an overweening faith in the qualities of the French soldier in immediately convened a *cour martiale* and the twenty men were sentenced to death *pour l'exemple*. The general then had second thoughts and decided that only the four corporals should be shot. The firing squad was less than enthusiastic about their duty and two of the corporals had to be finished off by the officer in charge just before a reprieve arrived from the repentant divisional commander. This sort of summary (and morale sapping) justice was, on the whole, absent from the methods of the British Army.

As they marched to war in 1914, the soldiers of the British and French Armies formed an extraordinary contrast. The British, having learnt the lessons of the Boer War, wore khaki uniforms, with peaked caps and puttees. They carried Lee-Enfield rifles, which served the British Army faithfully through two world wars. Officers wore similar drab uniforms enlivened by riding boots and swords. The cavalry fought dismounted and led more often than rode their horses into battle,

although in 1914 all regiments carried lances as well as carbines. French infantry were picturesque by comparison. They wore a dark blue overcoat buttoned back to assist movement and bright red baggy trousers; headgear was the *kepi*. Light infantry had light blue rather than dark blue jackets with yellow collars. Infantry of the line carried a considerable load: two leather ammunition pouches and on each man's back a leather pack with rigid frames in which could be found a blanket, a waterproof cape, a spare pair of boots and a mess tin. Cavalrymen carried a Lebel carbine (often contemptuously referred to as a 'pop gun') and a wooden lance. Dragoons wore brass helmets covered by canvas or felt to stop reflections from the sun giving away their position, and dangling from this was a horse hair plume.

French troops man a machine-gun post while their officer looks out for the enemy. This position probably saw little action; it would have been very vulnerable to artillery.

158

Left *A French colonel leads his men into the attack in Champagne, bearing the regimental flag aloft himself while his men have fixed bayonets.*
Right *British soldiers in defensive positions near Armentières during the winter of 1914–15. By now the fronts had become static as trench warfare took over.*

The heavy cavalry, the cuirassiers, wore Napoleonic style breastplates that made them look rather like medieval knights. Cavalry officers usually came from the best families of France and they got on famously with their British colleagues.

For an account of what life was like in the BEF during the early months of the war, we turn to Frank Richards of the 2nd Battalion, Royal Welch Fusiliers. He arrived in Rouen on 10 August and was billeted in a convent. The men of the BEF had been cheered in the streets and they were in a mood to celebrate. Lord Kitchener, the Secretary of War, had issued each man with a pamphlet warning him of the dangers of French wine and women but nobody seems to have taken much notice of this and Richards and his friends tried to order some wine at a cafe, though they failed because they could not speak French. Later, men wishing to enjoy the favours of a young lady would trade rare commodities; jam was especially prized. '*Mademoiselle, confiture?*' was a common refrain. 'In those days,' Richards commented, 'British soldiers could get anything they wanted and were welcome everywhere!'

On 24 August, when a million Germans invaded France, the men of the BEF were forced to retreat from the position they had reached near Mons and marched in a south-westerly direction. Richards was unaware that he was retreating; the officers told their men that they were luring the Germans into a trap. This was a time of short rations and blistered feet as the men struggled back to Le Cateau bearing their 50 pound (20 kilo) packs in the sweltering August heat. Richards's unit reached Le Cateau at midnight and as the men had marched through the whole of the previous day and night most fell asleep in the market square, though others tried to find some bread. At 4 a.m. they began marching again. This time they were told to leave their greatcoats and packs behind, taking only their rifles, ammunition, spare socks and 'iron rations' – four army biscuits, a tin of bully beef and a small portion of tea and sugar. Churches in passing villages were converted into hospitals to receive casualties from the battle raging behind them at Le Cateau.

Richards became separated from his unit and had to spend the next week trying to find it. He rejoined the Welch Fusiliers just in time to take part in the Battle of the Marne. During all this time the men were not allowed to take off the equipment they carried: one who did was sentenced to No 1 Field Punishment for twenty-eight days. Regardless of their excuses, the colonel paraded all stragglers (including several young officers) and meted out extra route marches as punishment. 'Our colonel was very strict, but a good soldier', commented Richards.

By this time, many men had lost their caps and their uniforms were wearing out. Some men covered their heads with knotted handkerchiefs and obtained pairs of civilian trousers. Looting from French civilians was common, but if caught an offender would be punished with death. Richards had his first dose of trench warfare near Fromelles when his battalion was ordered to 'dig in' by platoons, with a forty-yard gap between each platoon's trench. These trenches were designed as fighting shelters, not homes; they were only breast high and there were no sandbags. At night half the men took it in turns to stand guard for an hour at a time, but wet ground prevented the others from sleeping during their 'rest' periods. When the Germans finally attacked Richards' platoon, they learnt some of the principles of trench warfare: not least that ten defenders could easily hold off fifty Germans. When they began shelling the trenches instead, the British 'stayed put' and kept their heads down and were ready for the German infantry when they attacked again. After the final attack, the British would bury their dead; to be chosen as a member of the burial detail was not considered gruesome as we might think, but a piece of good luck. Useful pieces of uniforms and personal belongings taken from the dead were regarded as 'perks'; German bodies were also plundered, their spiked helmets being especially prized – they made useful latrine buckets until proper trenches could be dug.

The unhealthiness of life in the trenches was made

The First World War

The fighting man of the First World War faced a battlefield which was larger and more deadly than ever before, and on which men had to use outmoded offensive tactics in the face of an ever-improving technology of defence.

The basic problem is shown from two angles in the two diagrams *a* and *b*. First of all, an attacking force could be bombarded deep in its own rear areas by artillery such as the British 6-inch howitzer (shown below with the other main field weapons of the Allied forces). Once an attack had started then the problems of bringing up reinforcements was greatly complicated by the ability of defensive artillery to pour fire into rear concentrations. Those troops involved in the attack itself had other problems, however. The accuracy of field artillery such as the French 75mm and the British 13-pounder combined with the use of proper time fuses allowed defending gunners to use air bursting shrapnel. This could be concentrated on one segment of the battlefield through which attacking infantry moved at their peril. Even when they had crossed this belt of destruction they were very vulnerable, because machine guns, often firing on fixed lines to create a mutually supporting pattern of death, could be relied on up to ranges of 2000 yards. The Vickers and Hotchkiss were two of the most effective of these weapons; the Vickers, with its water-cooled barrel, was very reliable and could sustain fire for long periods. Finally, the attackers would come under fire from magazine rifles such as the French Lebel and British Lee-Enfield while still 800 yards from their objective. Such a combination of obstacles ensured that warfare was static; and it also ensured that casualties were extremely high.

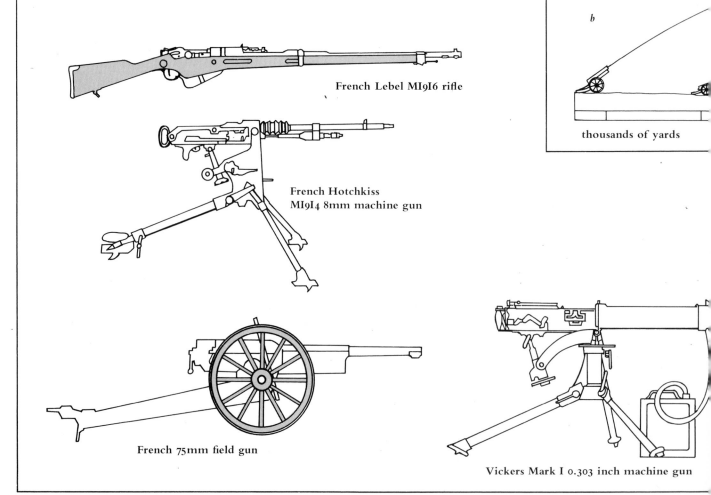

a

6-inch howitzer

13-
arti

b

thousands of yards

French Lebel MI9I6 rifle

French Hotchkiss MI9I4 8mm machine gun

French 75mm field gun

Vickers Mark I 0.303 inch machine gun

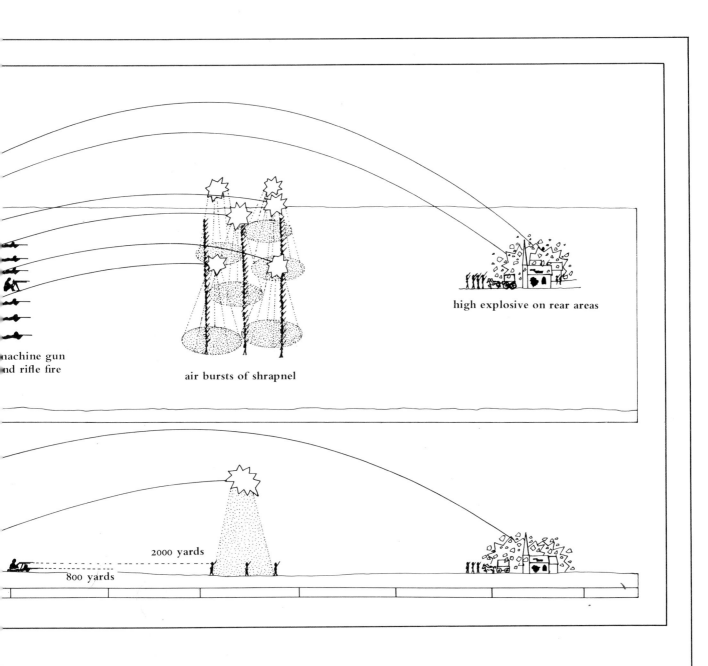

machine gun
and rifle fire

air bursts of shrapnel

high explosive on rear areas

2000 yards

800 yards

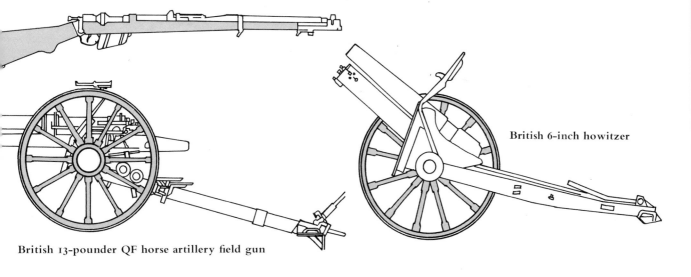

British Lee-Enfield Mark III 0.303 inch rifle

British 6-inch howitzer

British 13-pounder QF horse artillery field gun

The glory of the French cavalry was to no avail on battlefields dominated by shrapnel and the machine gun, but the armour-clad cuirassiers left for the front with high hopes, and great pride in regimental tradition.

worse by the fact that no fires were permitted; all food and drink was cold, and no water could be boiled. By the time Richards' unit was relieved on 15 November, the men were covered with lice and caked with mud, since none of them had been able to wash or shave for nearly a month. The carnage on the Aisne and at Ypres had taken its toll. None of the eighty-four battalions of the BEF had retained more than half of their strength by the middle of November 1914. Only two officers, two corporals and twenty-seven men survived of the original 1000 men of the 1st Battalion, Queen's Royal (West Surrey) Regiment, and these were mainly from the transport and cooking details. A total of seventy-five of the eighty-four battalions had been reduced to 300 officers and men or less in a period of four months. Britain's old professional army had been destroyed, although in the process it proved itself a magnificent fighting force.

The fighting had been ferocious. Here are the views of a German reserve lieutenant in Flanders:

> In the first few engagements our battalion was reduced to about half . . . We were at once struck with the great energy with which their infantry defended itself when driven back, and by the determined efforts made by it at night to recover lost ground . . . The main strength of the British undoubtedly lies in the defence and in the use of ground. Their nerves undoubtedly react better than those of the Germans, and their sporting instincts render them easier than our men to train in shooting, and in the use of ground and patrolling. The hardiness of their infantry was very apparent at [the First Battle of] Ypres. The shelter trenches were so well constructed that they could not be discovered with the naked eye . . . My own observation shows me that the British are excellent at patrol work, which I cannot say of our men.

A German chaplain, too, testified that the British were 'cold-blooded and tough and defend themselves even when their trenches are taken, quite different to the French'. Even the Kaiser had to admit in 1915 that the 1st Corps of the BEF which had fought at Ypres was the 'best in the world'.

French soldiers went to war in a pitch of enthusiasm, confidently believing that France would get her *revanche* for the humiliating defeats at the hands of the Prussians in 1870–71. These high hopes were dashed. French soldiers were excellent fighting material: they were intelligent, self-reliant, patriotic and though they could be cynical, responded well to courageous leaderships. Morale was excellent, and the soldiers marched into battle cheerful and gay. On the march the conscripts soon got fit – they had no choice, either they bore the gruelling strain or they died of exhaustion. But the French tactics undercut the courage of the soldiers and presented the Germans with marvellous targets. An observer noted of a typical French attack in August 1914:

> Whenever the French infantry advanced, the whole front is at once covered with shrapnel and the unfortunate men are knocked over like rabbits. They are very brave and advance time after time to the charge through appalling fire, but so far it has been of no avail. No one could live through the fire that is concentrated on them. The officers are splendid; they advance 20 yards ahead of their men as calmly as though on parade, but so far I have not seen one of them get more than 50 yards without being knocked over.

A French corporal likened the ubiquitous German machine gun fire to 'a coffee mill, *tac-tac-tac*. The bullets stream past; it is an infernal uproar. With each bullet I think: That one is for me'. In the Battle of the Frontiers in August 1914 French casualties were immense; the total losses in the last five months of 1914 were 754,000. Captain Grasset, a company commander, recorded an action at Virton in the Ardennes:

> The wounded offered a truly impressive sight. Sometimes they would stand up bloody and horrible-looking and admit bursts of gunfire. They ran aimlessly around, arms stretched out before them, eyes staring at the ground, turning round and round until, hit by fresh bullets they would stop and fall heavily.
> Heart-rending cries, agonising appeals and humble groans were interrupted with the sinister howling of projectiles. Furious contortions told of strong bodies refusing to give up life.
> One man was trying to replace his bloody dangling hand to his shattered wrist. Another ran from the line holding his bowels falling out of his belly through his tattered clothes. Before long a bullet struck him down. . . . I was wondering in my anxiety, whether we were going to lose all our men on the spot.

Under these conditions the excellent French cavalry was utterly useless. Their equipment quickly wore out, and in addition the French cavalry trooper had little idea of how to look after his horse. Unlike his British counterpart he did not dismount at every opportunity, and so French horses quickly got sore backs and were utterly exhausted in a few weeks; the wastage rate was so high that one regiment from every French cavalry division had to be withdrawn by the second week of

August. Examples of French dragoons charging German positions across muddy fields with their lances on foot were recorded in October 1914 – a futile act of great courage. The French cavalry was on the whole much superior to the German, but the Germans relied more on infantry support and rarely ventured out into the open against the French. 'If only they would get on their horses and fight. They are supposed to be cavalry aren't they?' was a common complaint among French cavalry officers. Major-General Spears, the British liaison officer with the French Army gave a graphic description of the French cavalry during the retreat to the Marne as he drove into the village of Craonne on 5 August 1914:

> An endless column of motionless cavalry completely blocked the road. The great towering cuirassiers, clumsy and massive in helmet and breast plates, sat impassive on their horses. Not a man dismounted . . . A gust of wind animated the horsetail plumes that hung down each man's back, then the long steel clad column was still again. There was something stoical about these men that was very striking. . . .

The early battles of 1914 came as a profound shock to both Britain and France. The war was not 'over by Christmas'. Britain had to raise new conscript armies from scratch, so that by 1918 there were ten times more British divisions in France than had been present in 1914. The men in these new armies were younger and less experienced; the average age of an infantryman in 1914 had been thirty, by 1918 it was down to about twenty. France's exaggerated romantic notions about war had been shattered. Gone were the gaudy uniforms of 1914. In 1915 the dark blue coats and red trousers were replaced by the *horizon bleu* which acquired a grubby sameness in the trenches of Champagne. French soldiers now acquired a stoical rather than dashing appearance; they smoked pipes and grew beards and came to be called *poilus* – 'hairy ones'. Their courage was still great but the organization of the French Army was inferior to that of the British forces. The French medical service was notoriously incompetent – to be picked up by a French stretcher party was a death sentence to wounded *poilus*. Provisions for leave were scandalously inadequate. A French private was entitled to seven days' leave every four months, and this was subject to cancellation at any time; neither were arrangements made to transport the troops home.

Under the strain of great casualties and disgust at the false optimism about victory 'just around the corner' spread by the General Staff, the grievances of the *poilus* became more vocal. Discontent broke out into open mutiny in 1917 and for a time mortally endangered the Allied cause, while in Britain disillusion with the war was undermining the morale both on the front and at home where the war effort was disrupted by a wave of strikes. By the end of 1915 the vigour and determination which distinguished both the French and the British forces in their own ways had in the face of the realities of trench warfare been replaced by something very different.

14
THE GROSSDEUTSCHLAND DIVISION

The German Army of the Second World War was one of the most effective of modern times. It was ultimately defeated by vastly stronger forces, but its units were regarded as the best, in general, of the war. Yet this marvellous fighting machine was almost wilfully blind to the enormous crimes of the Nazi Party. It is difficult not to admire the sheer fighting ability of the German Army, which must be counted (together with perhaps the Roman Legions of the first two centuries AD and the armies of Napoleon from 1805 to 1812) among the best European armies ever; but at the same times the seeming refusal to become aware of the horrors being committed in the name of Germany must loom large in any assessment of this army.

The nature of the German fighting man is perhaps best examined through the Grossdeutschland Division – a crack formation which grew from regimental to divisional and finally to corps status, and endured some of the heaviest fighting on the Eastern Front.

The martial tradition in Prussia had, of course, been strong since Frederick the Great. And the empire founded with the unification of Germany in 1871 had a profoundly military ethos. After defeat in 1918, however, the Treaty of Versailles restricted the German Army to 100,000 men and 4000 officers. To prevent reserves of trained men being built up, officers and men were only allowed to serve twenty-five and twelve years respectively, and the general staff and the War Academy were abolished so that Germany could no longer prepare for war.

The deeply-engrained tradition and pride of the German Army Command, however, and a general feeling among the general public in Germany that the Versailles terms were too harsh, meant that it was inevitable that an attempt would be made to circumvent the stipulations of the treaty.

General Hans von Seekt, who headed the German Army was a prime mover in this development. Maintaining utmost secrecy Seekt managed to come to an arrangement with the Soviet Union and by the Treaty of

Two members of the Grossdeutschland, showing the dress uniform of the early period of the war, contrasted with the combat gear worn on the Eastern Front.

Rapallo, the Soviet Union agreed to co-operate closely with Germany in the manufacture of aircraft and German troops trained in Russian tanks. A secret training programme was instituted to increase the number of divisions from seven to twenty-one, and attempts were made to create thirty-nine 'border guard divisions'. The General Staff was revived by calling it the Military Office, and vast stores of weapons were made, pilots were trained and short service training programmes introduced for the reserves in the universities and boys' clubs. By the late 1920s the main military clauses of the Versailles Treaty had been successfully avoided and a foundation for a programme of rearmament had been established which was accelerated openly by the Nazis after 1933.

Seekt's *Reichswehr* was a true élite. Because of its small size, competition for the available military posts was intense, and the best men got the jobs. Seekt determined that it should be a *Fuehrerheer*, an army of leaders; his guiding principle was that each man should be trained so that 'he would be capable of the next higher step in case of war'. So colonels were trained to assume higher command; junior officers were to become colonels and majors; the NCOs would become junior officers and the private soldiers the NCOs. When in January 1933 Hitler came to power he determined openly to flout the Versailles Treaty and return the German Army to its former size and prestige. German soldiers were taught that the new *Wehrmacht* 'is the arms bearer of the German people. It defends the German Reich and fatherland, its people which are united by national socialism, and its *Lebensraum*'. The 'community of field grey' temporarily disrupted by the Versailles Treaty had been restored.

In 1933 Hitler had been shown three platoons of partly mechanized soldiers in training. They consisted of motor cycles, anti-tank guns, and light tanks co-operating with armoured cars. 'That's what I need!' he announced and thus was born the 'mailed fist' of the German Army – the armoured and mobile divisions. The creation of an armoured force is a complex undertaking. The habit and techniques of mobile warfare can only be taught after a prolonged period of training. In the 1930s Germany, despite the covert efforts under the Treaty of Rapallo, totally lacked a pool of mechanically minded men and tanks were few in number; specialized infantry and artillery carriers were conspicuous by their absence. Germany possessed next to nothing except the ideas of a few far-sighted men, notably Heinz Guderian, who pioneered the new tactics of armoured warfare.

Germany's lightning victories in the first year of the war led to the widespread myth that Germany was well prepared for war in 1939, straining at the leash to set out on the path of conquest. Yet German rearmament had been hurried and was haphazard in quality. The first Panzer divisions were formed in 1935, and by 1940 Guderian had nine at his disposal, with a total of 2574 tanks, but they were mostly equipped with Panzer Mark I tanks, a training vehicle armed only with machine guns. Panzer Mark IIs were a slight improvement on this with 30mm armour and a 2cm gun. The 'workhorse' of the Panzer Divisions, the Mark IV, only began to replace these inadequate vehicles in 1940. The role of propa-

ganda in spreading this fictitious idea of Germany's military strength was crucial. Foreigners were impressed by military parades and the Nuremburg rallies. They greatly overestimated the number of German troops under arms. The reality was somewhat different: in October 1939 after the Polish campaign, there were only sufficient munitions to supply one third of the *Wehrmacht* for four weeks. The Panzer divisions were not ready to take the field again until April 1940.

The cadre of the tank crews came largely from the old *Reichswehr*, and up to forty per cent of the officers came from cavalry regiments. The remainder came from the Panzer schools opened throughout Germany, like the one at Wunstorf, where recruits came from a variety of social backgrounds. It was said that they were easier to train than the old *Reichswehr* soldiers. The men were selected by merit. The *Panzertruppen* felt and acted like an élite. They were given special black uniforms and a large black beret to distinguish them in appearance

Left *German Panzer troops in Russia in 1941. The Mark II tanks with which they are equipped were, however, becoming a liability by 1941: they were no match for the Russian T34.*
Above *A parade of Sturmgeschutz III self-propelled guns. The white helmet was the divisional marking.*

In mechanized warfare the ideal training arrangement was to teach one member of the tank crew the trades of all the others. There were three main trades, driver, gunner and radio operator. But it takes a long time and considerable expense to train men to double up in this way, so Panzer commanders were often content that their men should only learn two of the tank trades. The early efficiency of the Panzer divisions sprang from the thoroughness of their training and the intelligence of their recruits. Also, they had received 'dress rehearsals' for war with the *Anschluss* of 1938 and the occupation of Czechoslovakia in March 1939. When they went to war in September 1939, the Panzer divisions consisted of two tank regiments, each of two battalions totalling approximately 560 tanks, a brigade of panzer grenadiers carried in lorries, engineers and anti-tank guns. The total complement was about 13,000 men.

The Grossdeutschland grew out of the *Wach Regiment Berlin*, a guard regiment made up of seconded men from other regiments. It formed guards of honour for visiting dignitaries and had the highest drill standards. Drill, 'spit and polish' and the barrack square dominated its ex-istence and every Saturday it practised the famous 'goose step'.

In April 1939 Hitler signed the decree transforming the *Wach Regiment* into the Grossdeutschland, an élite regiment that would represent the embodiment of the German warrior spirit. Recruits would be taken from all over Germany and were carefully selected. The virtues of hardship, renunciation, sacrifice, courage, obedience and manliness were to be impressed on every recruit; the end product it was hoped would be physically, mentally and morally fit for the rigours of war.

There was a conscious effort to link the Grossdeutschland to Germany's military past – the name itself bestowed a special status on the regiment since its meaning was highly emotive. Reaching back to the ambitions of Bismarck and the unification of Germany, Hitler revitalised the term in a way even the 'Iron Chancellor's' vision had not encompassed. In addition, the highly trained, almost mechanical professionalism of the Grossdeutschland, borne out of years of drill and parade in its days as the Wach Regiment Berlin reached back even further to the rigid genius of Frederick the Great, and accounted in part for the Grossdeutschland's extraordinary tenacity even in the face of overwhelming odds. The way this vigorous training was translated into action is eloquently demonstrated in just one sentence of an eyewitness account of the crossing of the Meuse in 1940: 'The groups make their way forward as if they were on a parade ground'.

A special uniform was designed for this unit which added white piping to the standard field grey, but when Germany went to war in September 1939 there was not time for this robe to be issued. As a special mark of distinction the Grossdeutschland regiment was authorized to wear a cuff band with the name of their unit woven in against a black background but even these were not universally distributed – the only ubiquitous marking was the shoulder strap with the initials GD entwined. When the regiment was expanded into a division in 1942, all Grossdeutschland vehicles were given special markings: a single, unframed, white steel helmet stencilled on to the left front bumper.

The Regiment Grossdeutschland's first campaign was in France and the Low Countries in the spring of 1940. As infantry supporting the armoured formations, they swept through Luxembourg and Belgium and then rested before joining Guderian's daring attempt to cross the River Meuse into France which attempted to outflank the Maginot Line. Summarized in *Germany's Elite Panzer Force: Grossdeutschland* by James Lucas (Macdonald and Jane's, 1978) is the account of the eyewitness already mentioned, Lieutenant von Coubiere (2nd Battalion Grossdeutschland Regiment) of the crossing of the Meuse. He wrote on 13 May:

> We have a long night march behind us, and now towards midday the sun shines hot and mercilessly. The companies of 2nd Battalion are resting near their camouflaged vehicles, sleeping and strengthening themselves for the battle which lies ahead or writing letters home.

The fighting as they approached the river was not difficult. The 2nd Battalion's objective was a strongpoint called Point 247. The battalion fought its way through Sedan and was temporarily held up by a bunker. 'Hand grenades smoke the enemy out', wrote von Coubiere. 'The French stand with their backs to the wall and raise their hands. They had been told that the Germans would shoot all Frenchmen caught in pill boxes'. The advance continued with all speed, determination and efficiency. The opposition was weak, and the French infantry showed little of their customary dash. The crossing of the River Meuse showed the enterprising and self-reliant German soldier at his best:

> There can be no pause; the objective must be taken in daylight. Then the road will be clear for the Panzers to reach into the rear areas of the Maginot Line. The enemy must be given no rest . . . The second pill box is taken and we can make out the anti-tank gun position. New assault troops are brought forward, the machine gun group fights down the riflemen's positions on the left and enables the company to move forward . . . Once again the Grenadiers move out. They climb the slope, through deeply cratered countryside, cross deep barbed-wire barriers until the French open defensive fire from behind a ridge. Machine gun and machine pistols send out their death-bringing bullets. Hand grenades explode; nobody pays any attention to the enemy fire. There is not time to

> stop. The leading troops are already in the enemy's position. Close combat fighting, hand to hand, with a wild swing the attack is driven forward.
>
> Point 247 is ours. The way to the south has been opened!

Grossdeutschland then proceeded to cross the Meuse itself. A sergeant of engineers continued with an account of how he crossed the Meuse. Crawling through ruined buildings, the sergeant and his party crept into their assault craft, and four men squeezed into boats designed to hold three. They were ordered to throw away all excessive equipment, including their spades as there would not be time to dig in once the river had been crossed. At this point the Meuse was 80m (90 yards) wide and the crossing was as exciting as it was dangerous – in fact it was considered a defensive line that the French generals had confidently predicted would hold out for at least a week. After their triumph on the Meuse, victory followed victory for the German forces and on 17 June 1940 the French accepted the inevitable and sued for peace. The armistice allowed the Grossdeutschland to rest and re-equip, but not for long as Hitler began preparations for the invasion of the Soviet Union.

The professional skill of the German Army was dissipated and eventually destroyed when Hitler invaded the Soviet Union, for the rolling steppes of Russia proved to be the graveyard of the *Wehrmacht*. In June 1942, the Grossdeutschland, which in France had served as the mobile infantry component of the armoured units, was redesignated a motorised infantry division, with its own tanks and an increased complement of support weapons. The Eastern Front was for the Grossdeutschland a scene of triumph and disaster. It fought there for almost four years and it was there that it grew from a regiment into a division and then into a Panzer Corps. But the awesome level of attrition sustained during these years of ferocious fighting wore out the Grossdeutschland completely, so that by 1945 it was all but a skeleton of its former self.

On the face of it Hitler's decision to send in the Grossdeutschland made sense. Their elite training and efficient organization gave them several advantages which made their deployment attractive in all situations, good or bad. Whatever happened, Hitler knew he could rely on their unstinting loyalty; this was just as well since there were many occasions when orders from HQ conflicted with the military sense of those commanders faced with the realities of the Eastern Front.

On 22 June 1941 Operation 'Barbarossa' commenced. Over three million Axis troops poured across the Russian frontier, the Panzer divisions driving deep into

Above right German mobile troops advancing in the summer of 1942, crossing the River Don.
Right Piles of frozen Russian corpses. The German units inflicted enormous losses on the Red Army during the last six months of 1941; tactical superiority was manifest in every encounter and the Soviet forces only just survived.

Soviet territory. Despite brilliant initial success due to superior German tactics and great German expertise in the methods of fighting mobile warfare, the sheer scale of the operation proved too much for the *Wehrmacht* and in October the tank forces were halted by the autumn rains which turned the Russian steppe into quagmire. The mud, known as *rasputitsa* by the Russians, became so thick that it could even suck the soldiers' boots off. Yet even summer proved to have unforeseen perils. The dust of the steppes got into German engines because they were not fitted with a simple dust filter and shortened their lives. Most units had at least half their transport out of service at any one time. Those that were in service were hampered by lack of fuel – one armoured assault could only be mounted by syphoning off petrol from all the other non-armoured vehicles.

By late October the weather began to turn cold and the ground hardened to the satisfaction of the German generals who thought they could recommence the offensive. But the drop in temperature was so great – night temperatures reaching minus forty-five degrees centigrade – that the poorly equipped *Wehrmacht* was

Left German troops rush up a light infantry gun during fighting in the outskirts of a Russian town. Such scenes were repeated countless times from 1941 to 1945.
Below With rifles and automatic weapons at the ready, heavily-laden German infantry advance towards Red Army positions in the autumn of 1942.

unable to cope. Engines had to be left running all the time through lack of anti-freeze, automatic guns jammed in combat and guns did not recoil properly after firing. The German troops suffered as they had not been equipped for a long campaign and had no winter clothing. Thousands died of exposure, and many lost their limbs to the surgeon's knife. One infantry battalion of the 24th Corps had 800 cases of frostbite in one day – eight times the number of casualties inflicted by the Russians. The Field Kitchens could not rise to the challenge. Food was poor and hardly ever hot. The soldiers were left to their own devices but there was little for them to find following the scorched earth policy of the Red Army.

As food, fuel and ammunition were given priority by the transport services, only a minimum quantity of winter clothing was moved up to the front. The men of the Grossdeutschland were very lucky if they received one winter lined coat for every five men. As distribution of the clothing proved difficult, the stocks that arrived were purloined by the men working in the rear echelons, while the men in the line were left to freeze where they stood or huddled in isolated farm houses. This hardly added to the popularity of the support services who were looked upon as leeches living off the blood of the fighting soldiers. In these extreme temperatures even slight wounds turned gangrenous as the men were under-nourished from lack of food and enfeebled by the cold. Neither were they able to wash or change their uniforms and the whole army was riddled with lice and vermin. Such clothes as they did own were continually wet from the snow and mud and offered little protection. As well as having to face these hellish conditions the nerves of soldiers in the Grossdeutschland were plagued by the sniping of Siberian riflemen who were acclimatized to Arctic conditions and excellently equipped. They were mounted on skis and moved rapidly, sweeping into the rear areas and shooting up the troops' billets; they would lie waiting for hours to get in a single shot.

But the Grossdeutschland survived the Russian winter and in April 1942 was raised in status to a Panzer Grenadier division, taking part in the great battles around Kharkov in the autumn of 1942. At this time the German divisions were still equipped with the Panzer Mk IV. This was a reliable tank armed with a 7.5cm gun which was up-graded from low to high velocity in 1942. Although the Mk IV had been continually improved it was inferior to the Russian T-34, though by the autumn and winter of 1942 the new Tiger tanks were being introduced into the Panzer Divisions. The Tiger was an exceptionally heavy tank, weighed 55 tonnes and was given a powerful 8.8cm gun. The summer of 1943 saw the arrival of the new Panther tank which embodied the sloped-armour design of the T-34. Although suffering from mechanical teething troubles the Panther, with its mobility, good armour and superb high-velocity 7.5cm gun, was the outstanding German tank of the war and was feared by the Allied forces accordingly.

A typical tank battle was as follows. Men cooped up in tanks had less scope to show their fear than men fighting

General Manteuffel, commander of the Grossdeutschland, (on the left) confers with a major on his staff. They have both been awarded the Knight's Cross.

in the open. They could not run away and the main way out, through the turret, was blocked by the tank commander. He usually only gave the order to abandon the tank once it had been severely hit. The tank commander would choose his target and give the necessary order. The gunner turned the dial with his left hand to bring the enemy machine into focus and give the range, while his right hand would move the elevating mechanism. Once sure of his target he would drop his hand, take off the safety catch, give the details to the commander and wait for the order to fire. The commander who might be standing up in the turret, depending on the intensity of fire, would give this when he was ready; once the target was hit he would search for another. When the Soviet tanks had been driven from the field, the Panzer Grenadiers would climb into their regimental transports and the advance would continue. This was sometimes easier said than done as Soviet attacks came without respite and with great resolution; their resources in tanks were so great that material losses hardly counted in Soviet calculations – so long as the position was taken.

One of the German Army's great strengths had always been the speed at which commanders were able to regroup for battle. In the fact of practically suicidal

German Armoured Warfare

The basic tactics of the German mobile and armoured forces during the Second World War are shown in diagrams *a* and *b*. After preliminary (but often short-lived) bombardment – including the use of close-support aircraft such as the Junkers Ju 87 'Stuka' to cut enemy communications – the attack along the line would commence, with the main weight being concentrated on one narrow point, to break-through with armoured forces. Once through, the tanks would push ahead, bypassing enemy strongpoints (or perhaps attacking them from the rear), leaving the infantry to mop up. These tactics could only work with adaptable troops confident in their ability, their methods and their equipment. The basic attacking formation was as above right: the blunt wedge. The tanks would have been on a wider space than shown here, with larger gaps between the vehicles. This formation gave the maximum scope for mutual support, while probing for gaps in the enemy defences. Six of the most important vehicles used by the Grossdeutschland in the later years of the war are shown below right. The Mark IV was the workhorse; an adaptable tank, upgunned and used throughout the war. The Sturmgeschutz III was cheaper to produce than a tank because it had no revolving turret. Troops would be carried on the SdKfz 251; but the pride of the armoured divisions was the Panther tank, which had the most penetrative gun and well designed sloping armour. The Tiger, by contrast, had a less effective anti-tank capability and was very short on range and manoeuvrability; it was better for defence than attack. Finally, the division would have mobile support artillery such as the *Wespe*, equipped with a 10.5cm howitzer.

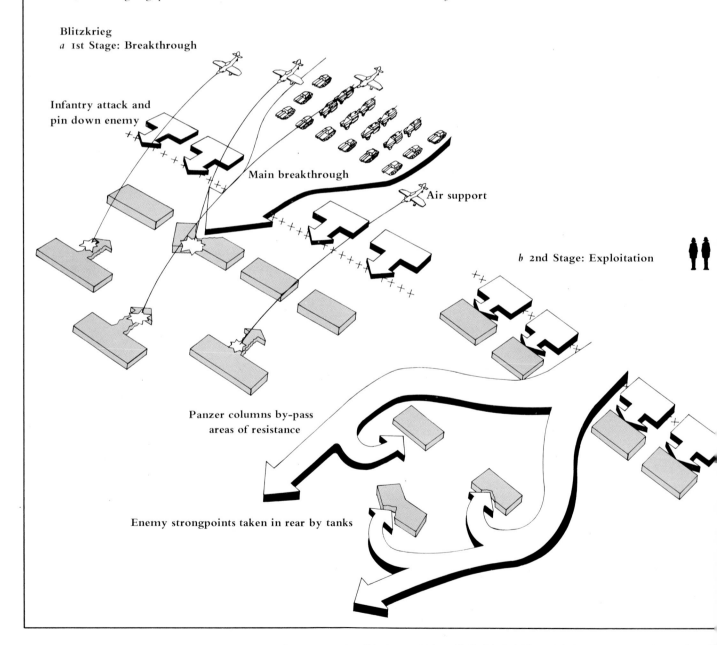

Blitzkrieg
a **1st Stage: Breakthrough**

Infantry attack and pin down enemy

Main breakthrough

Air support

b **2nd Stage: Exploitation**

Panzer columns by-pass areas of resistance

Enemy strongpoints taken in rear by tanks

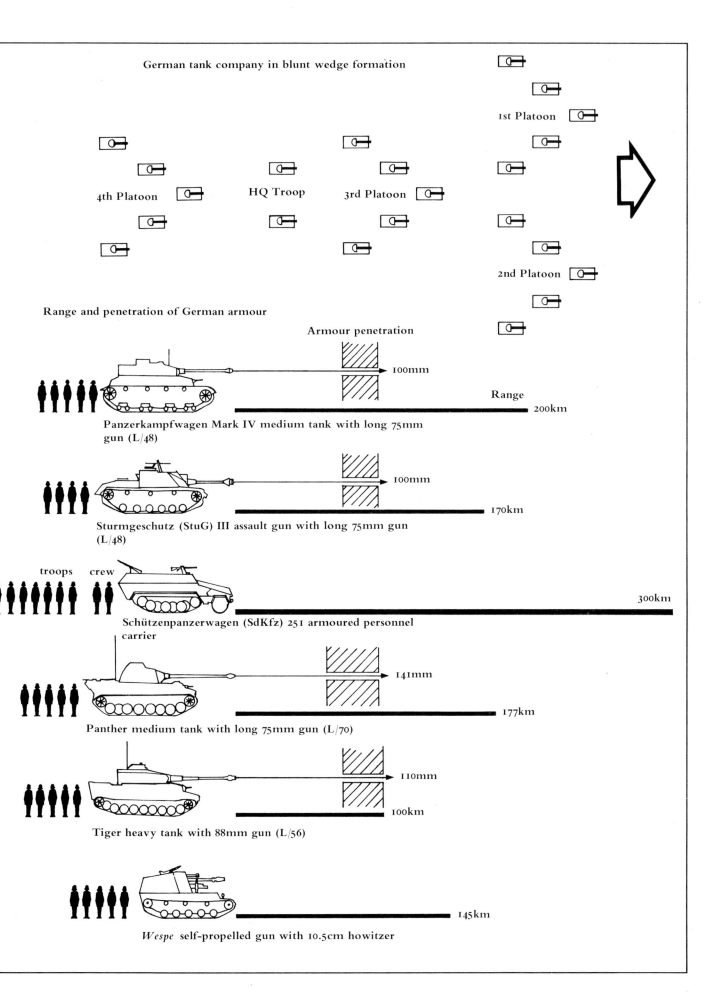

German tank company in blunt wedge formation

1st Platoon

4th Platoon   HQ Troop   3rd Platoon

2nd Platoon

Range and penetration of German armour

Armour penetration

100mm

Range
200km

Panzerkampfwagen Mark IV medium tank with long 75mm
gun (L/48)

100mm

170km

Sturmgeschutz (StuG) III assault gun with long 75mm gun
(L/48)

troops   crew

300km

Schützenpanzerwagen (SdKfz) 251 armoured personnel
carrier

141mm

177km

Panther medium tank with long 75mm gun (L/70)

110mm

100km

Tiger heavy tank with 88mm gun (L/56)

145km

Wespe self-propelled gun with 10.5cm howitzer

Soviet offensives, it was just as well for the German infantry that the Grossdeutschland Panzers could be called in, as they were at Rzhev in September 1942, at literally two hours' notice.

The fighting at this time was extremely uncertain, alternating between sniping and short sharp battles often reduced to hand-to-hand fighting with bayonets, knives or sharpened trenching tools. The wounded could not be removed fast enough; casualties increased because they were simply trampled in the mud. At one point the whole Grossdeutschland division was reduced to the size of a weak regiment; at another just two battalions' worth.

Badly mauled in the fighting for Kursk (Operation 'Citadel' in July 1943, where battle degenerated into the slogging matches characteristic of the First World War) the Grossdeutschland was allowed to refit in Romania in the autumn of 1944. The intensity of the fighting had whittled away the veterans of 1940 and the flow of replacements now consisted of boys of seventeen who were sent into the line almost as soon as they completed their basic training. Gone were the days when recruits were carefully selected; the martial strength of Germany was nearing exhaustion. But the Grossdeutschland still gave a good account of itself. In one five-day stretch of fighting, the Grossdeutschland Division destroyed seventy Russian tanks, shot down nineteen aircraft, captured forty-five guns and destroyed thirty-six AT guns.

But the vast size of the theatre of war and the numerical and material weakness of the Germans left them with a great feeling of helplessness and futility. The Russians advanced like a flood tide and they could live where soldiers from Western Europe would starve.

The Red Army's great offensive of 1944 had thrown the Wehrmacht back to the frontiers of Germany and the Grossdeutschland was transferred to East Prussia in a last ditch attempt to halt the Russian tide. It was thrown into a desperate attempt to relieve the beleagured Army Group North cut off in the Courland Peninsula in Latvia. This was a forlorn hope and only contributed to the further wearing down of the division. Trapped in the minefields and marshes of Prussia the division lost men at a catastrophic rate. Men of what had been considered an elite were having to face in a typical day a constant round of bombardment, rocket projectiles, mortar bombs and machine gun fire. Soldiers, some of whom were teenage boys or middle-aged men well past the normal age for active service, were expected to advance through land with scarcely any cover and fight the enemy hand-to-hand; if they survived till nightfall there was no respite from night patrols, lukewarm food if they were lucky, and – the irony of life in water-logged trenches – little water that was fit to drink. At the end of the battle, in most battalions the only officer left was the battalion commander; whole platoons were led by corporals and companies were down to a strength of forty men. The Grossdeutschland seemed to be on its last legs.

But in 1945 Hitler decided that it was best to make the most of Germany's dwindling resources by equipping the elite units with the very best material, and filling their ranks with the pick of the remaining recruits. The Grossdeutschland was revitalized. It was amalgamated with the Brandenburg Division and turned into the Panzer Corps Grossdeutschland. It was characteristic of the Third Reich (and certainly of Hitler's mania for numbers and new formations) that as its resources and power declined it created more and more units that it could not supply. The battles in East Prussia from January to May were fought in the streets of German towns where numbers were at a premium and the more sophisticated weapons that the Germans distributed

The differing battlefields of the Eastern Front: from armoured warfare on the Steppes (above left) to street-fighting in bitter cold (above).
Below right *A lieutenant of the Grossdeutschland discussing operations during the Kursk offensive of 1943. The initials 'G D' are intertwined on his shoulder strap.*

among their remaining fighting formations were of little use. The ground lost or gained could be measured in yards; the Germans were stubborn in defence and the loss of every farm out-building and house was contested. Members of the division had become adept at scrounging small numbers of weapons, especially guns, and small teams operated as 'tankbusters', seeking out and destroying individual Russian tanks. There was no longer enough fuel to move their own surviving tanks.

Driven back to Pillau, the last remaining port in German hands, the remnants of the divisions were evacuated to Bornholm; pursued by the Russians and determined not to surrender to them, small groups made their way to Denmark.

Although its fighting career lasted only six years, the Grossdeutschland – from regiment to Panzer corps – won for itself a reputation worthy of any of the great fighting units of history. Hitler built up the Panzer forces as part of his strategy to restore German power; he over-extended them in his determination to hold on to his conquests. The Grossdeutschland played a crucial and loyal part in his plan.

175

15
THE COMMANDOS

The flag of the Commando Battle Honours, which is to be seen in Westminster Abbey, bears the names of thirty-eight operations of war in which, between 1940 and 1945, the Army Commandos took part. The Commando Roll of Honour Book, which is also housed in the Abbey, contains the names of 1706 Commando soldiers, Army and Royal Marine, who lost their lives during the 1939–45 War. Eight members of the Commandos were awarded the Victoria Cross, six of them posthumously, as well as numerous other decorations. It is a record that compares well with that of any other British regiment or corps, including such as the Brigade of Guards, and the Parachute Regiment, who bore the heat of the day in World War II.

The Commandos' peculiar fighting tradition derived from their inception during the 'backs-to-the-wall' period of 1940. The keynote was self-reliance: it meant that the Commando soldier went into action prepared to act on his own, to fight with no more than the weapons and ammunition he could carry, to live off the land and if his officers or NCOs should fall, be ready to carry on with the plan for which he had been briefed. And if he got back from a raid, that was a bonus.

The Army Commandos came into being in June 1940 at the moment when the British Expeditionary Force had to quit the field, by way of Dunkirk, and when the Allies, one after another, were being compelled to abandon the struggle. The German *Wehrmacht* was 'riding high', and there were those who thought it invincible. Others, however, remembering how the German Army had been worn down in World War I, were convinced, however illogically, that the same thing could be done again.

There was little sign at that stage of the war that the United States, far less the Soviet Union, would one day be ranged on the Allied side. But many people in the Armed Forces and in government, not the least Churchill himself, believed that while retrenchment was necessary, special forces could be put into use for reconnaissance, raids and perhaps even invasion.

Two commandos, one wearing the 'light raiding order' including plimsolls, and the other with the equipment for amphibious landings.

These ideas were not entirely novel. The research department of the general staff had been looking into the training and deployment of guerrilla fighters from 1938. Under the aegis of Major (later Colonel) Holland, all types of irregular operations were championed, and he encouraged the formation of independent companies to be used in co-ordination with main operations. However, the first independent companies were not used to best advantage in Norway (they were used as the main force and not just for specialist tasks), so it was fortunate that Dunkirk changed the political climate and indicated to those in power the importance of developing truly independent forces, specially trained and equipped. In particular, the new Prime Minister, Winston S. Churchill, wrote in a minute, dated 18 June 1940:

> What are the ideas of C.-in C., H.F., about Storm Troops? We have always set our faces against this idea, but the Germans certainly gained in the last war by adopting it, and this time it has been a leading cause of their victory. There ought to be at least twenty thousand Storm Troops or 'Leopards' drawn from existing units, ready to spring at the throat of any small landings or descents. These officers and men should be armed with the latest equipment, tommy guns, grenades, etc. . .

Meanwhile Lieutenant-Colonel Dudley Clarke, R.A., had conceived the idea of striking back at the enemy by means of raids, an inspiration from his boyhood experiences of the Boer War in South Africa. He sold the idea without much difficulty to Lieutenant-General Sir John Dill, on the Imperial General Staff.

On 23 June the first Commando raid was carried out, near Le Touquet. It caused a few casualties and Dudley Clarke, who had gone along as an observer was hit by a bullet, which nearly severed one of his ears. There were three main lessons learned from that first raid: the problem of co-ordinating secret operations with regular forces, the need to be able to identify friend from foe and the importance (and difficulty) of pinpoint navigation at night. All of these were incorporated into the planning of the new Commando units which were formed during the summer of 1940. It was also clear from this first foray that in Europe at least, there could be no question of using Commandos as invading troops because of the difficulty of landing support weapons. In the first instance therefore the Commandos were to be used for reconnaissance and sabotage, and also for reforging intelligence links broken when the British were driven from the continental mainland.

The original Commandos were supposed to be volunteers. Units were asked to submit lists of officers and men for Special Service, though it cannot be said that there was much guidance as to what this was supposed to involve. It was an axiom of the British Army of those days never to volunteer, yet there was no lack of volunteers for special service at this time.

The Commandos derived much of their tradition from the ideas prevalent in the units from which the volunteers were selected. Several of the senior officers came from the Guards, and brought with them – besides blanco and brasso – the idea that if a job is worth doing, it is worth doing properly. When in 1940 the ten original Commandos were formed into a Brigade, it was to an Irish Guardsman, Brigadier (later Major-General) Charles Haydon, that their overall training was entrusted, and a former Guards drill sergeant, Lieutenant-Colonel Charles Vaughan, formed the famous Commando Depot at Achnacarry in Inverness-shire.

Potential Commandos were naturally expected to have a remarkably high standard of both practical skills and physical fitness, both of which were then polished during training, as well as important new information being taught. The official circular calling for recruits asked for 'courage, physical endurance, initiative and resource, activity, marksmanship, self-reliance and an aggressive spirit towards the war'. It is significant that the moral and psychological requirements are mixed in indiscriminately with the physical and practical. It is not perhaps surprising, then, that colonels of regiments became resentful as the Commandos threatened to cream off their best men; some became openly obstructive during the process of releasing volunteers for the Special Forces. On the other hand, it has been suggested that some found the Commandos a good way of off-loading awkward elements from under their command.

Selection may have seemed, at least superficially, to be idiosyncratic and privileged. For instance, in No 3 Commando, paradoxically the first to be formed, the commanding officer was a captain in the Royal Artillery, John Durnford-Slater, who sprang at the stroke of the pen to the rank of lieutenant-colonel. He selected his own officers, who were summoned to a country house in Romsey to be interviewed. Durnford-Slater allotted three to each of his ten troops, and sent them off to interview volunteers. In H Troop of 3 Commando this meant that Captain V.T.C. de Crespigny, R.A.S.C. Lieutenant J. Smale (Lancashire Fusiliers) and Second Lieutenant P. Young (Bedfordshire & Hertfordshire Regiment) did the rounds of 4 Infantry Division in de Crespigny's sports car.

Commando recruits were often reservists or territorials, mostly in their mid-twenties, who had experience in the field and whose initiative and skill at arms were already proven. There was however an extremely stiff fitness test to pass as well; at least twenty-five per cent of volunteers failed it. In some cases, only about half the establishment of NCOs was taken: many officers preferred to let the leaders emerge and promote from within the ranks. Either way Commando troops had a much higher proportion of officers and NCOs to other ranks than regular formations.

The initial plan was to raise ten Commandos, each with ten fifty-man troops (the equivalent of infantry platoons), but the limiting effect of the age and fitness of suitable recruits as well as the unwieldy administrative arrangements meant that in fact the total number of men in special forces was under 4000. In the field administration proved not to be such a problem – the emphasis on flexibility and self-sufficiency meant Commando soldiers did not require the back-up of ancillary staff such as caterers.

By November 1942, after considerable reorganization and the formation of the special training depot in Scotland, the strength of a full Commando was honed down generally to 24 officers and 430 enlisted men, and each Commando had about six troops, one for heavy weapons and five small rifle troops, organized into two thirty-man sections. However these divisions were by no means strictly adhered to. Emphasis was put on the development of the individuality of each Commando, and the trust which formed a bond between officers and men was used as the foundation of discipline. After all, a Commando soldier was supposed to be able to carry out a mission even if he had lost contact with his officers or fellow soldiers.

When they formed in 1940, the Commandos did not live in camp or barracks, but all were given the privilege of civilian billets which they arranged for themselves. For this they were given a lodging allowance: 13s 4d (67p) a day for officers, or 6s 8d (33½p) for other ranks. In those days, a second-lieutenant's pay was 10s 6d (52½p) per day, so one of the first things a subaltern discovered about Special Service was that for once he was solvent – a very rare feeling for junior officers. In Plymouth in 1940, where 3 Commando was based, excellent billets could be rented for two guineas (£2.10p) a week. Men were left to make their own arrangements, and found landladies ready to provide them with almost all their personal needs, in more senses than one. In other cases men did not fare so well: one officer in Plymouth camped out for some weeks on a golf course after a dispute over his billet.

At first training was left almost entirely to the various troop commanders, all of whom had their own peculiar ideas, many of which owed more to the cinema of Clark Gable and Spencer Tracy or the novels of Dornford

Commando training was at first rather haphazard, but the emphasis on physical skills and weapon handling was constant. One of the most important skills was landing swiftly on an enemy-occupied coastline, and exercises such as that shown above right developed this; while hand to hand combat training (right) was always a feature.

Yates than to any army manual. Even so, a fairly typical programme had emerged by 1941. Mornings were usually taken up with weapon training and unarmed combat, afternoons with map and compass exercises. Weekly route marches and tri-weekly exercises involving the whole Commando plus a healthy lack of routine within this programme kept soldiers on their toes.

The idiosyncrasies of individual commanders, while fostering the ideals of initiative, action and endurance, were really too unstructured even for the independent forces, and it was in February 1942 that the Depot (Commando Basic Training Centre) was set up by Lieutenant-Colonel Charles Vaughan at Achnacarry. The training programme outlined above was formalized into a crash course of about five weeks, including specialist training in canoeing, parachuting or cliff climbing; this last proved popular as a leisure pursuit – there was little else for the men to do in their spare time at Achnacarry. The assault course at the depot was legendary and live ammunition was often used in training. Men also learned variously to march at seven miles an hour, eat rats and kill silently with a knife in single-handed combat.

In spite of Churchill's pious hopes in June 1940, it was some time before funds could be found to provide Commandos with 'the latest equipment'. Essentially equipment was light, partly for ease of movement, and partly from lack of resources. Speed being of the essence in raiding, Commando equipment did not include mines and defensive wire, so they were very vulnerable to tanks although later in the war they did sometimes have anti-tank mines strapped to their equipment. They used conventional infantry weapons, such as the Lee-Enfield No 4 Mark 1 and Mark 3 rifles, the M1 Carbine and Thompson sub-machine guns; mortars were used at close quarters and grenades were adapted to give short or long intervals before detonation. Commandos also used tracer bullets to demoralize the enemy.

In 1940, lack of trucks and suitable boats really slowed down Commando development, and although this improved somewhat with their status, transport was always a problem. Among the more unusual modes of transport adopted by Commandos were canoes, midget one-man submarines, West African-style dories for beach landings, Dutch pram dinghies and bicycles.

For all this, it must not be forgotten that Commandos were soldiers, they wore regulation uniform, and there was nothing unusually clandestine about their organization even though so many of their exploits depended on the utmost secrecy. Indeed there were many critics of the 'fun and games' of the Special Forces. The Army Council refused a special cap badge in 1941, so soldiers wore the badge of their regiment of origin, and the famous Green Beret was not introduced until late 1942. It is therefore a measure of the loyalty within Commando units that the ultimate punishment was RTU – 'return to regular unit'.

After some months of frustration, two of the Commandos, 3 and 4 got the action for which they yearned, taking part in the first raid on the Lofoten Islands, which won for the Commandos the Battle Honour, 'Norway, 1941'. Soon afterwards, Lieutenant Colonel R.E. Laycock sailed for the Middle East with three Commandos, 7, 8 and 11, which were grouped together as Layforce: 7 and 8 fought in Crete and 11 forced the passage of the Litani River in Syria. But Middle East Land Forces were too short-handed to keep up the strength of these three units, and so they were disbanded, though not before Lieutenant-Colonel Geoffrey Keyes had won the Victoria Cross in an attempt to raid Rommel's Headquarters.

However it has to be said that the Commandos, for all their high spirits and their ruthless training, had not done much by the beginning of 1941 to justify their existence, especially to those critics in Home Forces who were very ready to assert that the Commandos could not do anything which could not equally well be done by any infantry battalion. This was not true. Nearly all the battalions in the Home Forces had suffered more or less heavily in Norway, in France and Belgium, and had therefore had large numbers of replacements to train: men scarcely out of their recruit training. In those days practically all commanding officers were in their late forties, and almost all were too old to go raiding. The

men, were, of course, willing enough to face the normal hazards of modern warfare, but that is not to say that they would lightly undergo the hardships and dangers which it was thought – not without reason – would be a peculiar feature of Special Service. It was moreover evident from the outset that it would not be easy to select targets for raids, or, having done so, to carry them out. Consequently, special forces had to be selected for this work. The critics, of course, were not always placated by these arguments.

By 1943 attitudes had changed radically, however. Those who had previously resented the Commandos or refused to take them seriously, now recognized their true worth and readily acknowledged their unique contribution. They were indeed 'real proper chaps' as General Montgomery put it, on seeing the entire force of 4 Commando climbing the Brandy cliffs at St Ives in Cornwall. The sort of operation which earned them this accolade is exemplified by the Vaagso/Maaloy raid, carried out by 3 Commando in the winter of 1941. It proved to be a classic – one of those rare operations of war when all the main tactical groups succeeded in carrying out their tasks.

Late in November 1941, Lieutenant-Colonel Durnford-Slater was summoned from Largs to London by Lord Louis Mountbatten who had recently taken over from Sir Roger Keyes at Combined Operations Headquarters. He was shown very complete intelligence covering the Vaagso/Maaloy area of mid-Norway, and was to advise the Admiral as to whether No 3 Commando could carry out a raid there. Sör Vaagso is a small port, on Vaagso Island, not far from that point of the Norwegian coast nearest to the Shetland Islands. It was used by the German navy as a convoy forming-up place. The anchorage was defended by two coast defence batteries, some anti-aircraft guns, and two torpedo-tubes. The garrison was thought to number 200 German troops. There were German warships, destroyers and MTBs in the area. At Herdla, Stavanger and Trondheim, the Germans had in all twenty bombers and seventeen fighters, all within fairly easy range. Colonel Durnford-Slater lost no time in assuring Lord Mountbatten that No 3 Commando was capable of wiping out this garrison.

The Commando was reinforced by one and a half troops from No 2 Commando, divided into five groupings. A detachment of Norwegian soldiers, under Captain Martin Linge, was distributed between the various troops to act as guides and interpreters. The 6-inch cruiser HMS *Kenya* and four attendant destroyers were to accompany the party to pound the battery initially.

Though Durnford-Slater drafted the original plan and commanded the force put ashore, the overall force commander was Brigadier J.C. Haydon, DSO, OBE, who took passage aboard the *Kenya* along with the Naval Force Commander, Rear-Admiral H.M. Burrough, CB.

On 13 December 1941, No 3 Commando embarked at Gourock aboard *Prince Charles* and *Prince Leopold*, two

Belgian cross-channel steamers, which had been converted to carry assault landing-craft, and sailed for Scapa. On 24 December after a number of exercises the force sailed for the Shetlands, and had such a rough passage, that the raid had to be postponed for twenty-four hours. As darkness fell the force sailed once more. The weather was still rough, but now it was a following sea. No longer did the Princes wallow, with the wind on the port beam. With the submarine HMS *Tuna*, as a navigational beacon the force found the narrow entrance to Vaagsfjord within a minute of the pre-arranged time and began the run-in.

Major Robert Henriques, Haydon's brigade major, and a celebrated novelist, conjures up those first minutes of the raid.

> It was a very eerie sensation entering the fjord in absolute silence and very slowly. I wondered what was going to happen, for it seemed that the ship [*Kenya*] had lost her proper element, that she was no longer a free ship at sea. Occasionally I saw a little hut with a light burning in it and I wondered whether the light would be suddenly switched off, which would mean that the enemy had spotted us . . .
>
> It was most disturbing that there was so little left to do because everything had been done beforehand. We noted the time, exactly one minute late, that the landing craft were lowered and could just be seen through glasses, black beetles crawling in the shadow of the mountains up the black waters of the fjord. We heard our aircraft overhead and saw their welcome of heavy, familiar tracer fire rising quite slowly from the surrounding slopes. Our ship was moving very quietly towards the headland where we should come into sight of the battery, which ought by now to be expecting our arrival. As we nosed round the point, everyone was waiting for the order to 'open the line of fire', and get in first with a salvo.

At 08.48 hours that order was given and *Kenya* illuminated Maaloy with star shell for the benefit of the naval gunners and of the Hampdens flying in to drop their smoke bombs and screen the landing places of Groups 2 and 3. For nine and a quarter minutes salvos of *Kenya's* shells thundered into Maaloy and the islet, 250 yards square, practically disappeared in the cloud of smoke set up by some 450 six-inch shells.

The German battery – four Belgian 75s, quick-firing field-guns taken in 1940 – could easily, if undisturbed, have sunk at least half the landing craft, creeping head-on towards them. It may have fired a few rounds, for empty shell-cases were found in No 1 gun's position: but, except for the fire of the little AA gun, that was all. Without the covering fire of *Kenya* the LCAs of Group 3 would not have been able to reach Maaloy unscathed. As it was they took the island in eight minutes at a cost of two or three men slightly wounded. Perhaps a dozen of the German gunners had been killed in the bombardment, and a few in the assault. Captain Butziger and most of his men were taken in their bunker. The Hampdens, with their 60lb smoke bombs had persuaded them that

there was an air-raid, and they had not anticipated that the British would come by sea and air simultaneously. Maaloy was the strongest enemy position, and it was there that all had expected the stiffest resistance so the ease with which it was taken was fortunate.

At Halnoesvik Group 1 found only two Germans, both of whom were severely wounded by a Lance-Corporal with a Thompson sub-machine gun. Group 2 got ashore as planned with very little resistance from the Germans. Unfortunately at the last moment the armed trawler *Föhn*, lying in Ulvesund, put a long burst into one of the Hampdens and the bombardier let his smoke bomb go a second or so too soon. The 60lb phosphorous bomb fell into one of the 4 Troop's landing craft, and few indeed of the soldiers got ashore without injury or burns. It was a horrible episode, and at a stroke it robbed the assault group of a quarter of its force. The survivors pushed on undaunted as they had been trained to do, and prevented a group of Germans taking up their defensive position.

3 Troop had landed unopposed and worked towards the Church under spasmodic fire, which soon began to take its toll of casualties. Despite valiant efforts, the men were gradually brought to a standstill. Ignoring initial casualties, 4 Troop pushed forward through the din and smoke, wiping out successive enemy posts and snipers. Durnford-Slater, who caught a glimpse of the troop captain, Forrester, wrote later that he was 'throwing grenades into each house as he came to it and firing from the hip with his tommy gun. He looked wild and dangerous'. At length he came up against the German HQ in the Ulvesund Hotel and tried to storm it, but was killed in the attempt along with the Norwegian captain, Linge. Corporal White, however, managed to throw a grenade into the hotel, which, rather surprisingly, caught fire. With Giles, Forrester and Linge dead and all but one of the other officers hit the momentum had gone out of the attack of 3 and 4 Troops. The Colonel looked about for reinforcements. The first to appear were some of 2 Troop from Halnoesvik led with Hibernian verve by Lieutenant O'Flaherty, who got a footing in one of the warehouses on the waterfront, and, though slightly wounded, shot down the first man to resist him.

The colonel now ordered Captain Bradley to bring up as many men of his troop as could be spared from their demolitions tasks, but this produced few reinforcements. Brigadier Haydon was asked to send in the floating reserve, Group 4, but the wireless equipment of those days was unreliable and the message took a long time to get through. A summons to Maaloy brought an officer and eighteen men of 6 Troop, who met with some good fortune in getting the advance going again. Men of 2 and 6 Troops stormed two buildings, wiping out the garrisons. At the second, they met with stiff resistance from a brave handful of German infantrymen. Here

Above right Brigadier Peter Young instructs two of his men in camouflage techniques before an operation.
Right The fighting on Vaagso.

Commando Tactics

The commandos were designed to act as small raiding parties, fighting in an unconventional manner; a typical raid by night might be as shown below, where a group of men are landed on an enemy coastline, and with sufficient prior intelligence to use the terrain can mount a successful attack on enemy coastal defences, perhaps in support of a more general landing. For such tasks commandos needed specialist skills (to scale the cliff and remain concealed before the attack) and also a massive weight of firepower at close quarters. These qualities were certainly needed in March 1942, when commandos led the raid (opposite bottom) which destroyed the dry dock at Saint Nazaire (one of the few dry docks on the Atlantic coast able to take German battleships). HMS Campbeltown, loaded with explosives, rammed the dock gates (and its charges went up nine hours after); the commandos landed and did what damage they could. Fighting was so heavy that of a force of 600, nearly two-thirds were lost. But the results were extremely important, and few troops other than the commandos could have achieved them.

Typical commando weapons – of various types. The dagger became the commando symbol, and was essential for quiet killing. The Sten gun was the most common sub-machine gun in use. The one shown is fitted with a silencer. Below that is a Bren gun: a light machine gun of great reliability, very accurate when used for single shots. Finally, the PIAT (Projectile Infantry Anti-Tank) was for use against armoured vehicles or emplacements.

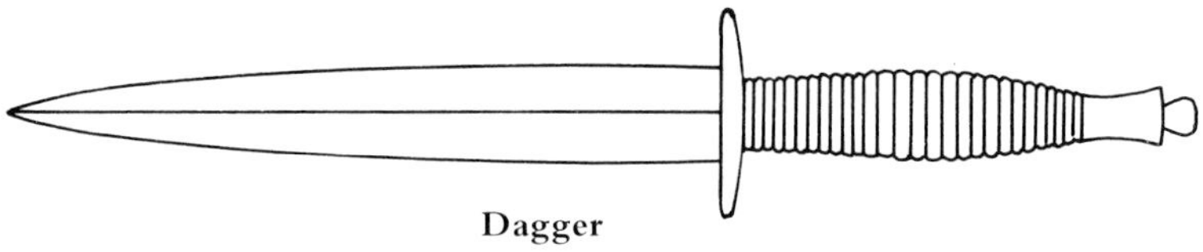

Dagger

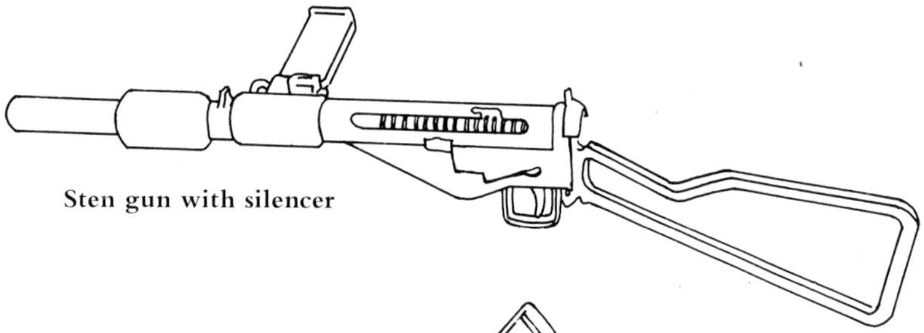

Sten gun with silencer

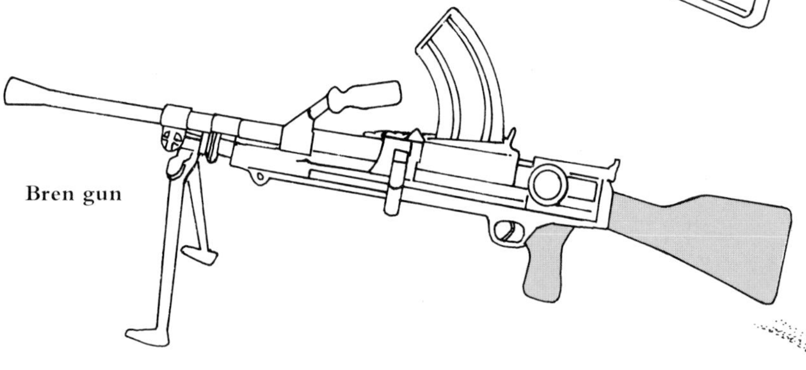

Bren gun

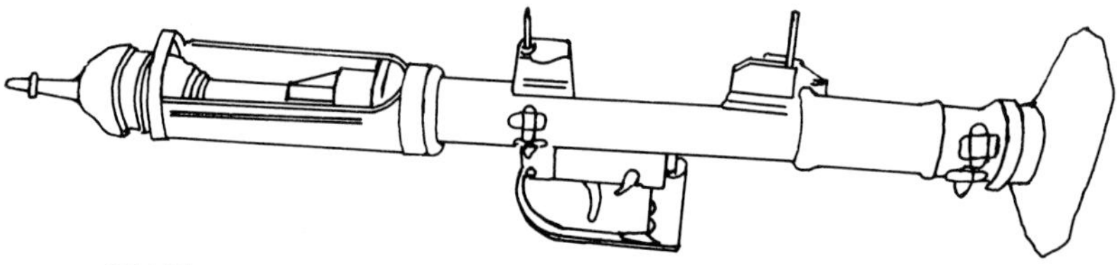

commandos land

PIAT

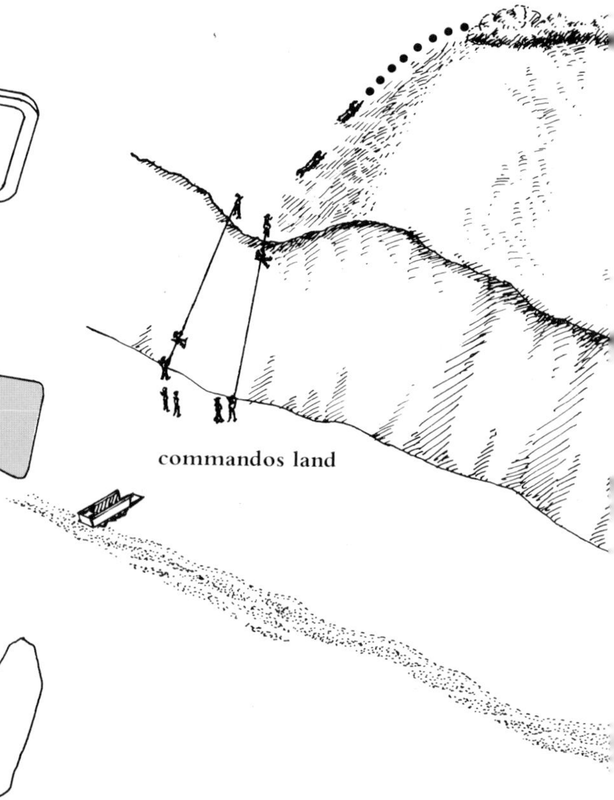

Hypothetical Command

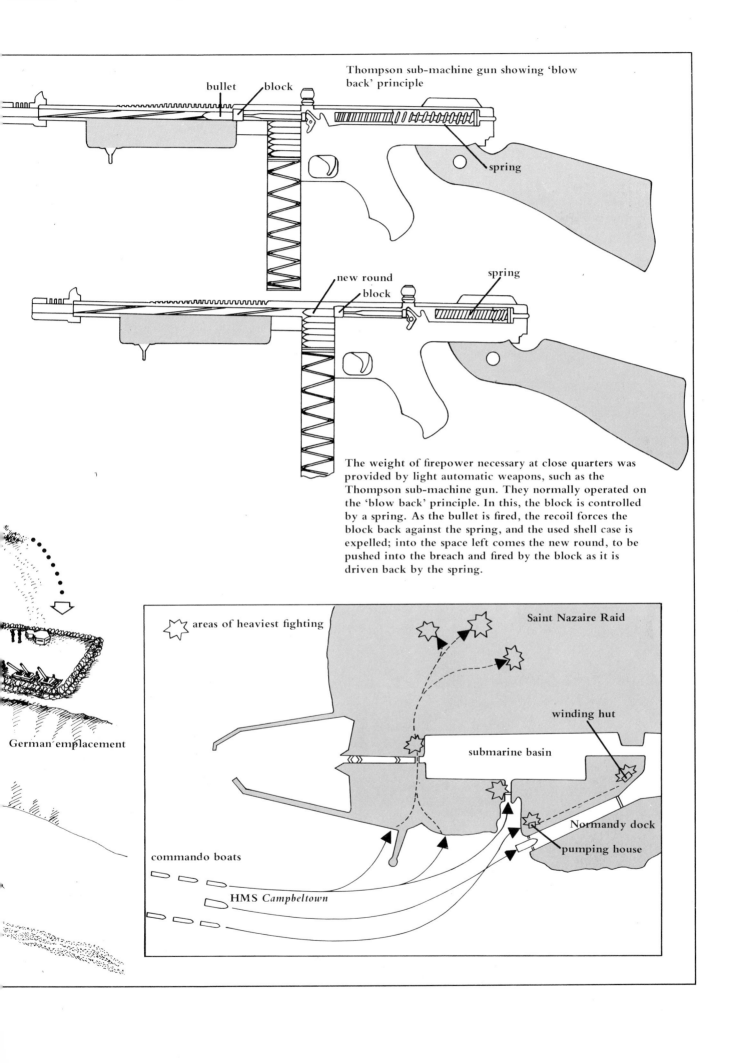

Thompson sub-machine gun showing 'blow back' principle

bullet

block

spring

new round

block

spring

The weight of firepower necessary at close quarters was provided by light automatic weapons, such as the Thompson sub-machine gun. They normally operated on the 'blow back' principle. In this, the block is controlled by a spring. As the bullet is fired, the recoil forces the block back against the spring, and the used shell case is expelled; into the space left comes the new round, to be pushed into the breach and fired by the block as it is driven back by the spring.

German emplacement

areas of heaviest fighting

Saint Nazaire Raid

winding hut

submarine basin

Normandy dock

pumping house

commando boats

HMS *Campbeltown*

O'Flaherty was gravely wounded, before the place was set on fire and its defenders compelled to emerge. Pushing on 6 Troop took the Myrestrand house, which proved to be the billet of the German commander, Major Schroeder, who was found upstairs, mortally wounded. His men had escaped. He and his men had sold their lives dearly, and when the army officers, including the chaplain, had fallen, the German harbourmaster, one Sebelin, and Stabsfeldwebel Lebrenz, a veteran of the 1940 campaign, had carried on the struggle from house to house and wall to wall. There were still a few holding out

when the time came to withdraw, and in almost the last incident of the day a lone German sailor emerged from an alleyway and threw a stick-grenade which wounded Durnford-Slater and both his runners before he was himself shot.

The withdrawal was conducted without great difficulty, and the British had hardly left the fjord when two companies of German infantry arrived by boat from Bergen. Next day Generalmajor Kurt Woytasch (181 Division) appeared to survey the damage. Eleven dead were found on Maaloy and thirty-five others were

missing. There were no survivors. Exactly how many had fallen in Sör Vaagso was not quite clear. The British had lost twenty killed and fifty-seven wounded, but as the Germans got no prisoners they were left with practically no idea how much damage they had done. Besides guns and a tank destroyed, a number of fish oil factories had been demolished. The Germans lost two armed trawlers, an armed tug and four steamers, and the majority of their crews, the total disposed of amounting to about 15,630 tons.

To the British the neat little victory at Vaagso came as a pleasant change after two years during which a patient people had been starved of successes. It came, too, as a happy omen for Lord Mountbatten's period as Chief of Combined Operations. Its effect on the enemy was successful too, and contributed to the diversion of thousands of German troops and naval personnel to the unnecessary defence of Norway.

Commando officers were, of course, supremely fortunate in having picked volunteers to lead. It has already been said that critics believed this deprived the line battalions of the best potential junior leaders. There is certainly an element of truth in this. On the other hand the Achnacarry training set a standard for the rest of the

Left Watching for German soldiers as a fuel dump goes up in flames in the harbour.
Below A wounded British officer is helped to the advanced dressing station by his men.

army, a standard, which every good commander, however much he might dislike the Commando idea, was determined to surpass. The spirit of emulation played an important part in the revival of the British Army after Dunkirk. It may be suggested that with the end of the raiding period it would have been as well to disband the Commandos, but in fact amphibious warfare continued until the very end of the war. The 3rd Commando Brigade, for example, did three major landings in January 1945, during the last Arakan campaign.

No one could suggest that the Commandos sprang to life as if Churchill, like a modern Cadmus, had simply sown the Dragon's teeth. The soldiers who stormed Vaagso with such dash, but also with some disregard for the finer tactical points of 'Fire and Movement', matured in Sicily and Italy into the veterans of Normandy. If that can be said of No 3 Commando the same is true of Nos 1 and 6 Commandos, who went from a hard apprenticeship in North Africa to perform with skill and determination in Burma and Normandy respectively.

After Agnone General Dempsey, no mean judge, told Brigadier Laycock: 'The men of No 3 are the finest body of soldiers I have seen anywhere.' Some of the Commandos were more fortunate than others in that they saw more action; and some formed a remarkably high opinion of themselves, an opinion encouraged by the press. But this opinion was not entirely unjustified, as they showed in a hundred fights between 1940 and 1945.

16
THE WARRIORS
OF BUSHIDO

The Japanese were latecomers to the race for overseas empire, but in the fifty years between 1895 and 1945 they made up for lost time. Formosa, the Kurile Islands and the Ryukyu Islands had been secured by 1904, and then the newly modernized Japanese forces took on the might of Czarist Russia at Port Arthur and Mukden and gained control over Korea, southern Sakhalin Island, and a considerable part of Manchuria. Having sided with the Allies in the First World War, Japan won a League of Nations mandate over the Marianas, Caroline, Palau and Marshall Islands in the South Pacific. The outbreak of fighting in southern Manchuria led to a full-scale (although undeclared) war between Japan and China in 1931, and in the following year the Japanese were able to

Left A Japanese tank officer, with samurai sword, and a typical Japanese infantryman.
Below The tradition of taking a constant offensive ran deep in the Japanese army. Here troops prepare to advance during the Philippine campaign of 1941.

establish the puppet state of Manchukuo; by 1937 they had captured Peking, Tientsin, Shanghai and Nanking as well, and in 1938 they seized Hankow and Canton. French Indochina came under their control in July 1941, and after their entry into the Second World War in December of that year they rapidly overcame every remaining Allied possession in the Far East; Singapore, Britain's 'Gibraltar of the Orient', garrisoned by 120,000 men, fell to half that number of Japanese invaders in less than seventy days. To those westerners who thought of all orientals as myopic, excitable midgets whose technology only extended to the manufacturing of cheap toys, the Japanese tide of conquest came as a rude surprise; it was a military record which could stand comparison with any of the great powers of the nineteenth or twentieth century.

As the Second World War progressed, however, startling reports of the Japanese soldier's methods of fighting began to reach the ears of the Allies; if he was not to be despised, he was nevertheless very different from

his western counterparts. European and American fighting men were not accustomed to rushing forward *en masse*, armed in some cases with sticks and rocks, against enemy artillery and machine guns; their officers did not normally disembowel themselves with swords if they lost a battle, nor were soldiers expected to blow themselves up with hand grenades to avoid being taken prisoner; aviators in western armies rarely chose to deliberately crash their aircraft, complete with bombs, into enemy ships, leaving no possibility of survival. All of these events took place in the Japanese army, not once or twice but hundreds of times, and they clearly showed the type of behaviour that was expected of the Japanese soldier. What was it that made a man throw his life away deliberately with little hope of military advantage to be gained by it, or persuaded a soldier to continue his fight alone, cut off from all military support, for more than twenty-five years, waiting in a jungle and refusing to believe that his country had lost the war?

The answers are *bushido* and *kodo* – 'the soldier's code' and 'the Imperial way'. *Bushido* was not merely a set of orders; it existed on a higher plane than that. Specifically, it was the ethical system of the old feudal class of *samurai*, the warrior caste of pre-modern Japan, which had been abolished in 1867. During World War II this ethic pervaded the entire army, including soldiers of peasant stock and officers from merchant families, as well as those descended from *samurai* forefathers. *Bushido* taught the virtues of absolute loyalty to a leader's commands, an austere way of life, honour, and courage. The earlier tenet of 'if it is right, the true man must be ready to die for it' had gradually been turned around to the claim, 'if you are ready to die for it, then it must be true'. The soldier was thereby compelled to prove the righteousness of his cause by dying for it, and death in battle was seen as the logical fulfillment of the soldier's career rather than a fate to be shunned. Because of this, a soldier must always fight to the last, as the Japanese war minister, General Araki, observed:

> Retreat and surrender are not permissible in our army . . . To become a captive of the enemy by surrendering after doing their best is regarded by foreign soldiers as acceptable conduct. But according to our traditional *bushido*, retreat and surrender constitute the greatest disgrace and are actions unbecoming to a Japanese soldier.

Fighting spirit was of far greater importance than military planning and armament, although these were by no means ignored, and failure was viewed as a lack of spirit on the part of the individual; he could redeem his reputation for courage only by enduring the agonies of *hara-kiri*, preferably by plunging his sword into his abdomen and drawing it across, thereby disembowling himself. This part of the code of *bushido* was actually incorporated in the army regulations during World War II, and any Japanese soldier who allowed himself to be captured was to be executed if he fell into Japanese hands again.

The other important ingredient in the Japanese

soldier's character was *kodo* – 'the Imperial way' – of absolute belief in the sacredness of the emperor and the destiny of Japan. The Home Islands of the Japanese people were created by the first gods and ruled in the beginning by the sun goddess; one of her great-great-grandsons, Jimmu, was the founder of the first human dynasty of Japanese emperors. Every successor of Jimmu, including the Emperor Hirohito who acceded to the throne in 1926, was therefore divine, and Japan's possession of this divine leadership was the guarantee of her own destiny to rule over the peoples of Asia. Just as every Japanese was related to the emperor, whose ordained role was to govern, so the divine plan called for the benighted peoples of Asia to acknowledge Japanese rule for their own good. Japanese society had always been rigidly heirarchical, and after the break-up of the feudal system in the late nineteenth century, the focus of loyalty was concentrated on the emperor alone; an imperial rescript of 1882 enjoined soldiers to remember that the command of a superior officer was morally equal to a command from the emperor himself. Given absolute loyalty, disdain for death, and belief in Japan's destiny on the part of the soldier, it was inevitable that the Japanese would triumph over their enemies according to the plan laid down by the gods.

The Japanese did not insist on direct rule over their conquered territories, but relied on puppet governments; they were thus able to present their struggle as a campaign to liberate Asians from rule by the western powers. The important thing was to secure Japanese access to raw materials, especially oil. A handbook presented to soldiers when embarking for the Singapore campaign claimed:

> The aim of the present war is the realization, first in the Far East, of His Majesty's august will and ideal that the peoples of the world should each be granted possession of their rightful homelands. To this end the countries of the Far East must plan a great coalition of East Asia, uniting their military resources, administering economically to each other's wants.

Hatred of the white race and belief in the policy of 'Asia for the Asians' profoundly influenced many Japanese soldiers during World War II; luckily, Japan's German and Italian allies did not have any Far East colonies at this time.

The aggressively military attitudes of *bushido* although accepted by the Japanese soldier, were not inherent in the population as a whole; there were plenty of eligible young men who tried to avoid army service and who looked upon the *samurai* code as a relic of barbarism. But all able-bodied adult males were liable for compulsory service, and they were prepared for this at an early age; semi-military instruction was introduced into the curriculum when the Japanese boy was eight years old. He

The conquerors of Singapore marching proudly through the streets of this bastion of the British Empire after the garrison had surrendered.

received further instruction under army officers during his secondary-school career, and if he left school early he was forced to attend part-time classes. At the age of nineteen he underwent physical examination and was placed in the appropriate group: if he was in good health and at least 1.52m (5 feet) tall he could be called up as required; if his eyesight or hearing was poor he went into the conscript reserve; otherwise he was placed in the local defence group, together with boys aged seventeen to nineteen. As the Second World War progressed, the standards for each class were gradually lowered, and eventually even Koreans and Formosans, normally regarded as fit only for labour battalions, were accepted for combat duties. Most of the conscripts for the army,

however, were of Japanese peasant stock, averaging about 1.6m (5 feet 3 inches) in height and weighing between 50–59kg (110–130lb); they were tough, healthy and used to hard work.

As Japan's military successes during the twenties and thirties piled one upon the other, her military leaders welcomed the system of conscription because it gave them the chance to instil the principles of *bushido* and *kodo* into the entire male population, and each new recruit underwent a rigorous three-month course of indoctrination to turn him into a fanatical warrior. But physical training, too, was important, and the recruit was forced to undertake prodigious feats such as an 80km (50 mile) route march with full pack followed by three circuits of a

191

Japanese Defensive Systems

The Japanese armed forces emphasised the primacy of the offensive, but after 1942 were fighting on the defensive almost everywhere in the Pacific theatre. A typical area for a small unit to defend would be a coral atoll – a string of islands barely breaking the surface atop a horseshoe-shaped coral reef, as in the inset. The basis of this defence would be one of the islets, on which an airstrip, some hangars, barracks and gun emplacements would be sited, within a defensive perimeter. The emplacements themselves (bottom) are interesting: they often had to be built out of coconut Logs, with coconut fibres providing a soft covering, because of the lack of loose earth or sand on the coral islands. The main defensive pill boxes were covered by machine-gun nests (opposite below).

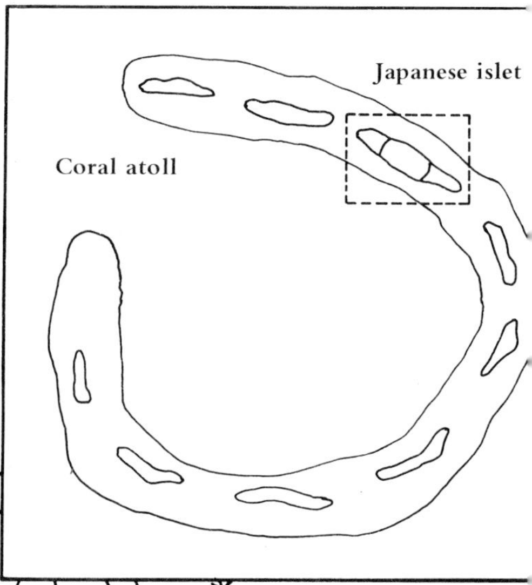

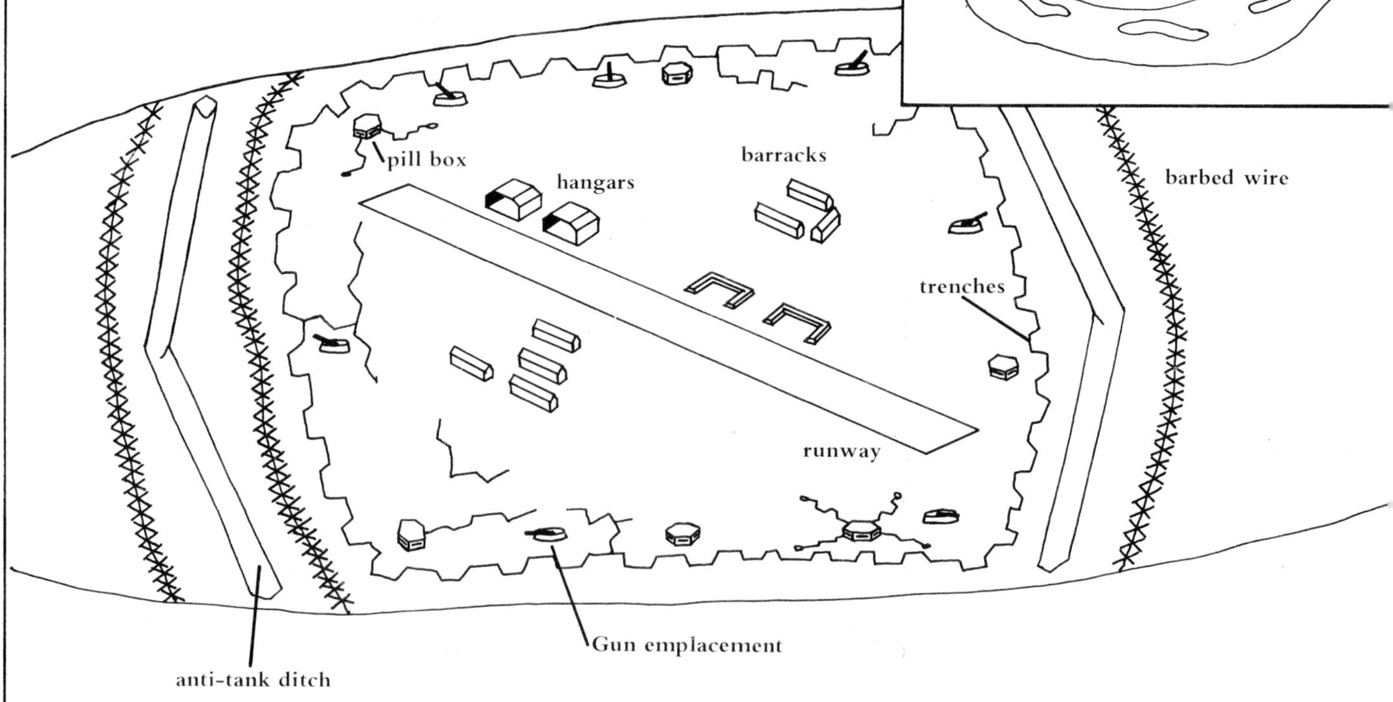

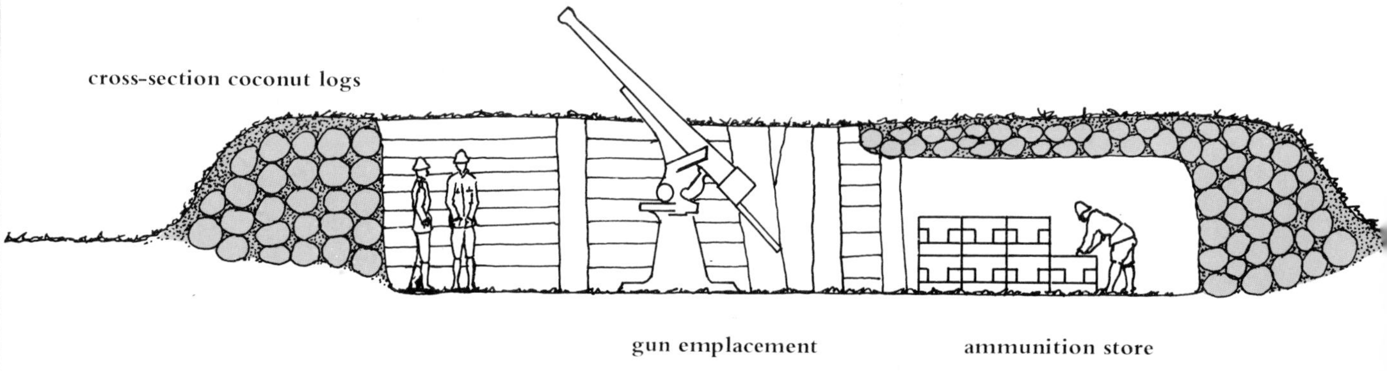

Coral atoll

Japanese islet

pill box

hangars

barracks

barbed wire

trenches

runway

anti-tank ditch

Gun emplacement

cross-section coconut logs

gun emplacement

ammunition store

Arisaka Model 1905 6.5mm rifle

Shiki Kikanju 1939 (Type 99) 7.7mm machine gun

The basic weapon of the Japanese infantry detachments was the Arisaka rifle (not a very efficient weapon) and close support was provided by the Type 99 air-cooled machine gun. The latter was produced in great numbers in the later years of the war, and provided the major firepower of an infantry platoon.

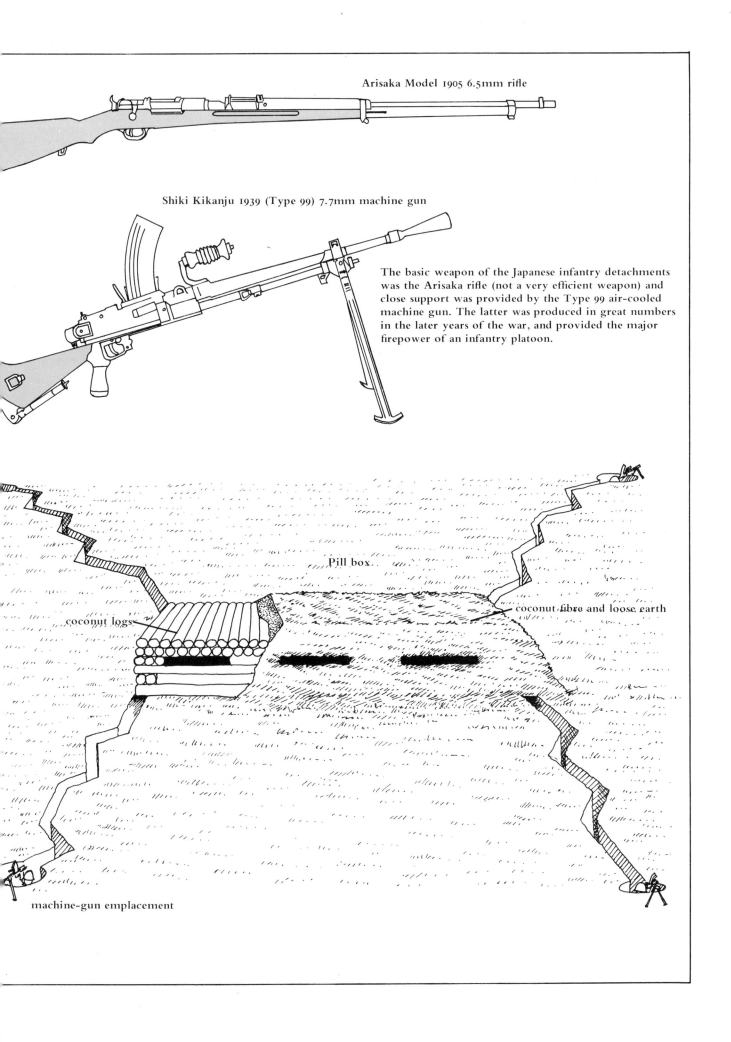

Pill box

coconut fibre and loose earth

coconut logs

machine-gun emplacement

field at the run. After one such ordeal an officer commented: 'They aren't nearly so exhausted as they think they are'; a good example of the emphasis on spiritual as opposed to physical resources. During field exercises the new soldier would be shot at with live ammunition to ensure realism. He would also receive special training in night operations and in movement through thick jungle, since both of these were known to be avoided by other armies whenever possible.

Barracks were not generally used in Japan; instead, the soldier was billeted in the local community. Since almost every family had a son who had been or was still in the army and was billeted somewhere, civilians generally accepted their enforced guests with a good grace. The soldier's pay was abysmal, but he was expected to keep his uniform in immaculate condition, if necessary by doing repairs himself. His rations were much the same as those he had eaten at home: unpolished (brown) rice (more nutritious than white rice), dried fish or beans, supplemented on occasions with chicken, pork, fruit and vegetables, and *saki* (rice wine) on special days.

The soldier benefited from various welfare programmes established by the military authorities, especially during the years of the depression in the 1930s; the army drew most of its support from peasants and workers, and adopted a paternalistic attitude towards them. Illness while on active service, however, was inexcusable. The soldier was expected to work or fight, and if he could do neither then he was not entitled to eat. More than likely, it was thought that the illness was his own fault – a sign of lack of spirit.

There were three types of officers in the Japanese army. Those destined for the highest ranks were the regulars, educated for leadership in military academies from an early age, and often sent as military attachés to foreign posts during the 1930s. The second sort were men risen from the ranks; they provided the majority of the lower-ranking officers on medium-service commissions, and were expected to do their jobs conscientiously without complaining. The last type were officers selected from the ranks of the conscripts to serve short commissions. A career as a regular officer was one of the most highly respected professions available in Japan, and unlike the private soldier the officer was well paid. He was, of course, expected to conduct himself according to the principles of *bushido* even more than the soldier was, and the most respected officers were those who best exemplified the warrior spirit; only a small proportion of the officer corps concerned itself with planning, logistics and technical questions. As an American war-time treatise of advice for men fighting the Japanese stated:

> The Japanese officer. . . is a magnificent leader of men.
> His weakness consists of his failing to remain master of a
> combat, as European officers do. He goes through with a
> battle rather than directs it. . . . The Japanese is more of a
> warrior than a military man. . . . The essential quality of
> the warrior is bravery; that of the military man,
> discipline.

The Japanese officer felt a close link with the men under his command, since all were fighting for the same reasons and with the same philosophy; at the same time, the officer had the moral authority of the emperor behind him and he expected that his men would follow him to the death without questioning.

For the Japanese officer, the sword was more than a badge of rank; if he was of *samurai* stock, it might be a family heirloom with 600 years of history behind it. The sword was carried everywhere, and often made the officers easy targets for Allied snipers. The *samurai* sword was a two-handed weapon with a hilt which was half as long as the blade itself, and was of no use in modern warfare; nevertheless, it was often discovered in the cockpits of crashed fighter planes. Private soldiers, too, revered their weapons in the tradition of the warrior; the day when the new soldier was issued with his rifle was marked with ceremony, during which the soldier would bow to his weapon before proudly grasping it for the first time.

Throughout his training the Japanese soldier had been urged always to take the offensive and maintain it with speed and determination. This was, of course, in accordance with the code of *bushido*, and it worked well during the ten years of fighting in the thinly defended, open countryside of China. Once the Japanese soldier came into contact with the enemy he had a great advantage, given his greater spirit and his often superior weapons. There was therefore no need for complicated battle plans, although the Japanese tried to envelop the enemy whenever possible, attacking the enemy's front with one column while sending another (and often stronger) around the flank to hit at the rear. The double-pincers movement was also used. The officers and unit headquarters were kept well to the front of the attack in order to direct the exploitation of weak points in the enemy's defences as soon as they developed. When the enemy had been crushed, the Japanese would immediately seize all available transport – including civilian vehicles, boats and bicycles – and rush forward deeper into enemy territory to maintain the momentum of the attack.

During the Second World War, the Japanese made particularly good use of bicycles, especially in Malaya; they were inexpensive, reasonably quick, and did not use up precious oil. The military planners ensured that all officers and men who could not be accommodated in motor vehicles were issued with bicycles, and a rapidly moving unit of Japanese cyclists surprised the headquarters of the British 6th Brigade in northern Malaya in December 1941. 'Even the long-legged Englishmen could not escape our troops on bicycles,' commented a Japanese colonel. 'Thanks to Britain's dear money spent on excellent paved roads, and to the cheap Japanese bicycles, the assault on Malaya was easy.'

The Japanese army won great victories in China, against the forces of Chiang Kai-shek.

Surprise was one of the favourite Japanese weapons, and they would sometimes attack with inferior numbers and without reconnaissance in order to confuse and discourage the enemy. Similarly, they relied on deep penetration by individuals or small groups into the enemy's rear areas – a very effective tactic for destroying morale. These tactics were, of course, dangerous, and the soldier was quite likely to be cut off; if this happened he would either make a suicide attack or lie in wait for a chance to use his grenades against the enemy before being picked off. Legends soon developed about the skill of Japanese snipers, but they were not in fact particularly good marksmen. On the other hand they were experts in the use of camouflage and would wait patiently for hours in a tree or undergrowth for the chance to account for as many enemy troops as possible before being killed.

Two excellent methods of surprising the enemy were attacks at night and attacks through jungle. Allied armies relied on roads to bring up their supplies and reinforcements, and they therefore regarded jungles and swamps as impassable; the Japanese were well-trained to take advantage of this, despite the loss of artillery and tank support (on which they did not place much emphasis, although they used it when available; Japanese tanks were at least ten years out of date). Night attacks were usually launched between midnight and 2 a.m., and allowed the Japanese to avoid accurate enemy fire (especially when they attacked through jungle); they could thereby attack targets which were too strong to be assaulted by day. If the first attack did not succeed, the Japanese would keep up the pressure on the enemy by blowing bugles or issuing threats through megaphones on the following nights, even if they did not actually intend to attack. These tactics succeeded in keeping inexperienced enemy troops on edge at first – for example during the struggle against the US Marines of Guadalcanal in 1942 – but the threats soon came to be regarded as a joke, and the Allies were able to prepare for the predictable Japanese night attacks and render them ineffective.

In fact, as the war progressed, Japanese casualties became proportionately higher with each campaign as the Allies learned to deal with their vigorous offensive tactics. Once the Allies got used to the idea that the Japanese placed little reliance on armour or artillery support, but emphasized the fighting spirit of the infantryman, they were able to bring their own massive firepower to bear directly on the attacking ground forces. As the losses mounted, however, the ideals of *bushido* did not change; if forced to occupy a defensive position for a time, the Japanese soldier was still likely to charge forward with his bayonet before his position could be overrun. Yet massive technical superiority did not ensure Allied forces an easy victory. Japanese soldiers in prepared positions – as on Iwo Jima and Okinawa – fought to the bitter end, causing huge American casualty lists for every position taken. Japanese soldiers had to be winkled out of every dug-out, often with flame-throwers, in a merciless round of close-quarter fighting.

As we have seen, capture by the enemy was the greatest disgrace that the Japanese soldier could endure, and this naturally coloured his view of Allied prisoners. As far as his comrades and family were concerned, the captured Japanese soldier was dead; he may have understood intellectually that this attitude was not shared by the Allies, but emotionally he hardly knew what to do with captured enemy soldiers. Japanese prisoner-of-war camps were generally run by very low-grade troops (shell-shocked, weak, sick, or otherwise deficient in warrior spirit) and since the prisoners were an embarrassment to the Japanese mind, they were left to rot.

If capture was the lowest depth to which the Japanese soldier could sink, death in battle was his high destiny.

The speed and lightness of bicycle transport was an important factor in the rapid conquest of Southeast Asia. Above Japanese soldiers on the attack in Penang, Malaya, during the conquest of the colony in 1942.
Right Japanese infantrymen in China issued with their staple ration – rice. These men are armed with Arisaka Type 38 rifles, rather out-dated by European standards, but the regulation army issue.

By dying for the Emperor he himself became something of a god. Twice a year the names of the fallen were inscribed on tablets and presented at the army shrine in Tokyo, and if any part of a soldier's body – his ashes or even his nail parings – was presented at the shrine during this ceremony, he would be deified. For this reason, soldiers setting out on campaign would be instructed by their officers to make their wills and to send them, together with their nail parings, to their families for safekeeping. This belief also explains why Japanese soldiers went to great lengths to recover their dead from battlefields, even at serious risk to themselves.

In the end, *bushido* was not enough to win the war, although it had often won battles in the early days. The fighting spirit of the warrior was no match for the massive firepower of the Allies, and the disparity in material resources between the two sides soon began to tell against the Japanese. In spite of this, the Japanese infantryman was a fearsome opponent right up until the end of the war, a fighting man of awesome bravery.

17
THE US MARINES

During the Second World War the marines earned themselves the reputation of being the toughest fighting men in the American armed forces. This reputation was based on a fine combat record (including some of the hardest fighting of the war), and exceptionally good public relations which ensured that the marines and their exploits were widely publicised. A key factor in the success of the US marines in the Pacific was their technological and logistical superiority over the Japanese. Armed with a wide range of personal weapons that included flame-throwers, phosphorus grenades, shot-guns and automatic rifles, the marine received back-up support that was the envy of other armies. Even on the battlefield the marine could call upon immediate artillery support from either shore or ship and direct air-strikes against enemy strongpoints.

Underlying the fighting image of the marines in the Second World War was the fact that they, unlike the other US armed forces, had one specific job to do and their own plan for doing it: the newly-developed doctrine of amphibious assault, which one British military historian has labelled 'the most far-reaching tactical innovation of the war'.

Although the marines had occasionally taken part in land-based campaigns when the US army needed the assistance of a trained force of professional fighting men – for example, as part of the American Expeditionary Force in France in 1918 – their primary responsibility was traditionally the defence of naval yards and the navy's advanced bases around the world. Following the British disaster at Gallipoli in 1915, most military theorists claimed that an amphibious assault against a well-defended beach protected by machine guns, barbed wire, artillery, and mines was no longer possible. Planners within the US Marine Corps, however, did not want to see the corps become obsolete, and they analyzed the Gallipoli campaign to find out what went wrong. New ideas were tested in the annual marine manoeuvres

Left A marine lieutenant and a private, the latter wearing combat clothing typical of Guadalcanal.
Right A group of marines with trophies in the Solomons. Their range of armament is clearly visible – heavy and light Browning machine guns and Garand carbines.

during the 1920s; these included 'combat loading' of vessels to ensure efficient delivery of supplies onto the beach, the organization of shore parties with the specific task of unloading the supplies and evacuating the wounded, and experimentation with new types of landing craft. Radio liaison parties were also detailed to each landing force to advise on the best use of naval gunfire and air support. Meanwhile, Japan's expansion during these years convinced American war planners that a campaign in the Pacific was a possibility which had to be planned for, and that US troops would have to attack enemy beaches whether they liked it or not. In 1933, therefore, the Fleet Marine Force was introduced as part of the US Marine Corps; its mission was to provide specialized landing forces trained in amphibious assault techniques, and to communicate its knowledge in these techniques to other US or allied armed forces when required. By 1941, on the eve of America's entry into the Second World War, the Fleet Marine Force included nearly 26,000 men – a third of the corps' entire strength.

The decision to establish the Fleet Marine Force was finally vindicated on 7 December 1941 when the Japanese struck at the Pearl Harbor naval base in the Hawaiian Islands, bringing war to the Pacific. Within

two days, the Japanese had also overrun the hopelessly outnumbered marine defence battalion on the island of Guam, and by the end of the month the island of Wake had also been lost; the Philippines fell in May 1942. Now the marines had the opportunity to test their amphibious assault techniques in action by recapturing the lost bases and bringing the war to Japan via her own naval outposts. The Allied High Command committed the US Army to the 'Germany First' plan, while the navy would have to serve in both theatres of the conflict, but it was in the struggle against Japan on the islands of the Pacific that the marines were to make their major contribution.

Traditionally, recruits for the marines were volunteers – and highly motivated ones at that. Publicity for the corps did not promise recruits that they would learn a useful trade, but that they would fight. Asked his reason for joining, one recruit replied: 'I wanted to be a marine because I had always heard that the Marine Corps was the toughest outfit going, and I felt that I was the toughest going.' The ethos was also reinforced by the nickname of 'Devil Dogs' (*Teufelhunden*) which had been bestowed on the marines in 1918 by the Germans, during the fighting in Bellau Wood in France. Legend also had it that the marine commander on Wake Island during the Japanese attack, when asked by headquarters what supplies he needed, radioed: 'Send us more Japs!' It was this devil-may-care attitude which the marine establishment promoted with remarkable success. The Marine

Corps occasionally accepted draftees, including 75,000 during the Second World War (out of a total strength at the end of the war of 485,000), and a percentage of the volunteers probably joined the marines rather than be drafted into the army, but the corps was able to obtain most of its manpower from enthusiastic volunteers.

After signing on, the marine volunteer was sent to his recruit depot for basic training – either at Parris Island, South Carolina or at San Diego, California. Although both depots had undergone large-scale building programmes since the beginning of the war in Europe, the flood of volunteers after Pearl Harbor caught them by surprise: there were 10,000 new recruits in December 1941 and 23,000 in January, instead of the pre-war average of 3000 per month. Training schedules had to be temporarily shortened to accommodate the flood, but by the spring of 1942 the standard course of seven or eight

Left *The aftermath of the vicious struggle to get the marines ashore at Tarawa: a beach littered with bodies and a knocked out Sherman tank in the foreground.*
Below *Men in jungle camouflage uniform, a pattern which was effective when the wearer was stationary, but tended to give away a man's position when he moved.*

weeks in 'boot camp' was restored. The first three weeks included plenty of drill, first without arms and then with them, together with the memorising of the corps' history and customs, a part of training that was, and is still given considerable prominence in the Marine Corps, as a means of building up a 'corporate identity' distinct from any other branch of service in the American forces. The latter included the learning of special Marine Corps idioms, such as 'brig', 'head', 'scuttlebutt' and 'skivvies'. The recruit also had to acquire the necessary skill in laying out his clothing and equipment at the foot of his bed for morning inspection; this was known as 'junk on the bunk'. After this came three weeks on the rifle range with the Garand M1 rifle or (preferably) the shorter and lighter M1 carbine; both were self-reloading weapons, strongly made and easy to operate, with eight-round magazines. The M1 Thompson sub-machine gun – the military version of the gangsters' 'Tommy gun' – was also used by marine infantrymen, and after its introduction in December 1942 the M3 sub-machine gun became widespread; a cheap, effective, all-metal weapon, it was known as the 'grease gun'. The last week or two in boot camp were spent in bayonet practice and instruction in guard duties and ceremonial parades.

During his weeks in boot camp and the following month in basic field training with his unit at Camp Lejeune, North Carolina, or the huge Camp Pendleton in southern California, the recruit was kept under surveillance to see if he might be eligible for an Officer Candidates' Class. The selection boards were allowed to choose up to one per cent of the recruits for this programme; they had to be aged between nineteen and thirty-five and physically fit, as well as showing natural aptitude. College graduates were often enlisted in the corps with the promise that they would be chosen for the class. Another source of new second-lieutenants for the corps was the Reserve Officers' Training Corps programme, in which students joined the marine reserves but were allowed to finish their studies before beginning active service. Field commissions were also given to outstanding NCOs, and some specialists needed for technical areas such as electronics were commissioned straight from civilian life. Relatively few of the marine

officers during the Second World War and afterwards came from the traditional source of élite officers, the US Naval Academy at Annapolis, Maryland, and the war also saw the end of the dominance of the officer class by the Old South – for whom, it had been said, the initials USMC stood for 'Useless Sons Made Comfortable'.

Despite the corps's need for strong and healthy men as recruits, one section of the population was deliberately excluded: in December 1941 there was not a single black American in the marines. The commandant of the corps stated in January 1942: 'There would be a definite loss of efficiency in the Marine Corps if we have to take Negroes,' but the die had already been cast by President Franklin D. Roosevelt in an executive order of June 1941: 'All departments of the government, including the

Burning out Japanese positions with flamethrowers was one of the best ways of clearing enemy bunkers of their courageous, last-ditch occupants.

Armed Forces, shall lead the way in erasing discrimination over color or race.' Blacks could no longer be fobbed off with the army or navy, and the training camp of the marines' 51st Defense Battalion, with black enlisted men and NCOs serving under white officers, was opened at Camp Lejeune in August 1942. Ironically, neither the 51st nor the similar 52nd Defense Battalion actually saw combat during the war, although they did serve in the Pacific; the black marines who ended up doing the fighting were the labour troops of the depot and ammunition companies and the officers' stewards, men who were not given the responsibility of combat but merely designed to provide support for the white men 'up front'.

Aside from the defence of the islands of Midway, Johnston, and Palmyra, the first major campaign of the Fleet Marine Force was the assault on Guadalcanal in the Solomon Islands on 7 August 1942. The naval guns opened up shortly after 6 a.m., and then the landing craft carrying the 19,000 men of the 1st Marine Division began moving towards the shore while carrier-launched planes patrolled overhead, ready for air strikes if needed. The landing, however, was unopposed, and did not therefore provide a test of the effectiveness of amphibious assault plans. But the Japanese response was not long in coming: bombers drove off the ships of the landing force on 8 August before all the supplies were unloaded, the marines being deprived of naval and air support, which did not improve their morale. Within a fortnight the Japanese 17th Army had been reinforced and began launching suicide attacks against the marine defence perimeter and the captured airfield, and the 'leathernecks' were introduced for the first time to the fury of mass *banzai* assaults. Here, machine guns were the marines' best defence, and they did their work well. During one Japanese attack, Sergeant 'Manila John' Basilone's unit cut down dozens of men in front of their two-gun emplacement – but then ran out of ammunition. Basilone sprinted through heavy enemy fire to obtain some more ammunition belts, but on his return he discovered that the machine guns to his right had been silenced. Taking some men with him and carrying one of his guns on his back, Basilone then ran over to the silenced position, killed the enemy infiltrators, and lay in the mud and torrential rain clearing the jammed guns and firing the belts he had originally brought for his own unit. By the end of the attack, Basilone had 'contributed in large measure to the annihilation of a Japanese regiment', as the citation for his Medal of Honor testified. The total number of enemy dead in this attack was 800, for a loss of thirty-four marines. The Japanese colonel leading the attack committed suicide.

Further Japanese infantry attacks against the marines on Guadalcanal took place throughout September and October, usually at night, but they were all beaten off by American machine guns, mortars, howitzers and grenades. Meanwhile, the first aircraft units had arrived at the captured airfield. The marine pilots of the fighters and dive-bombers of this 'Cactus Air Force', known as

'wing-wipers' by the infantrymen – who, in turn, were labelled 'gravel crunchers' – soon gained air control over the island and prevented the Japanese from bringing in further reinforcements. The Americans continued to build up their own forces; men of the 2nd Marine Division arrived in November, and army troops were committed early in 1943. The Japanese were slowly pushed back as the marines moved forward to Cape Esperance each day, having learned to deal with the enemy's night attacks; in the closing stages of the campaign, the marines were able to call for naval gunfire support in shelling Japanese coastal positions. On 9 February 1943 patrols moving up opposite coasts met at the tip of the island; the Japanese had disappeared, evacuated by a naval flotilla on the night of 7–8 February.

Guadalcanal was the first time the Americans had taken the offensive in the Second World War, and the 'invincible' Japanese had been beaten. The marines' combat efficiency against the enemy had been proven. As Major General Alexander Vandegrift, the commander of the 1st Marine Division, later wrote: 'We needed combat to tell us how effective our training, our doctrines, and our weapons had been. We tested them against the enemy, and we found that they worked. From that moment in 1942, the tide turned, and the Japanese never again advanced.' But the doctrine of amphibious assault had not yet been tested, and the marines also picked up some bad habits in the jungles of Guadalcanal: standards of personal hygiene and care of uniforms and equipment suffered in the tropical heat and humidity, and they stayed low throughout the Pacific campaign; some marines made a deliberate affectation of sloppy dress and beards. More dangerously, returning veterans sometimes advised trainees to 'throw away the rule book' when fighting the Japanese. The 'gung-ho' spirit that was so much a part of the marine myth discouraged caution and made heavy casualties almost inevitable. This unfortunate tendency persisted in the marine corps and led to the situation later in Vietnam where the construction of necessary field-fortification was dispensed with, such behaviour being considered detrimental to the aggressive marine tradition.

Throughout the rest of 1943 the marines took part in the clearing of the rest of the Solomon Islands chain in cooperation with the army, but General Vandegrift was eager to use the marines' specialized knowledge elsewhere and leave the army to eliminate the remaining Japanese in the Solomons. The marines got their chance in November 1943 during the campaign to capture the Gilbert Islands in the Central Pacific: the 2nd Marine Division was given the task of seizing Tarawa, a strongly defended atoll, by putting into practice the amphibious assault plans developed over the past twenty years.

The first ingredient in the assault plan was naval gunfire to 'soften up' the defences and force the enemy to 'keep his head down', and the navy's guns opened up early in the morning on 20 November, stopping on schedule when the infantrymen should have been hitting

Amphibious Warfare

The marines used a mastery of technology and organisation to make themselves experts in amphibious warfare. The initial approach to a Japanese-held island would be made by a task force comprising aircraft carriers, battleships (or heavy cruisers) as well as the troop transports and escort vessels. Aircraft from the carriers would begin the process of softening up the defences (diagram 1) and then the heavy guns of the battleships would continue the process while the troops entered the landing craft (2). Then the landing craft would go in, while the naval vessels carried on their covering bombardment (3).

The waves of landing craft would be arranged as in the diagram below. First came the LCAs (Landing Craft Assault) which carried 35 men. They were supported by amphtracks, armed with a 75mm howitzer. Heavier support was brought in from LCTs (Landing Craft Tanks) which could carry 5 Sherman tanks each.

The co-ordination of all these elements required great skill and small miscalculations – as during the landing on Tarawa – could have serious consequences.

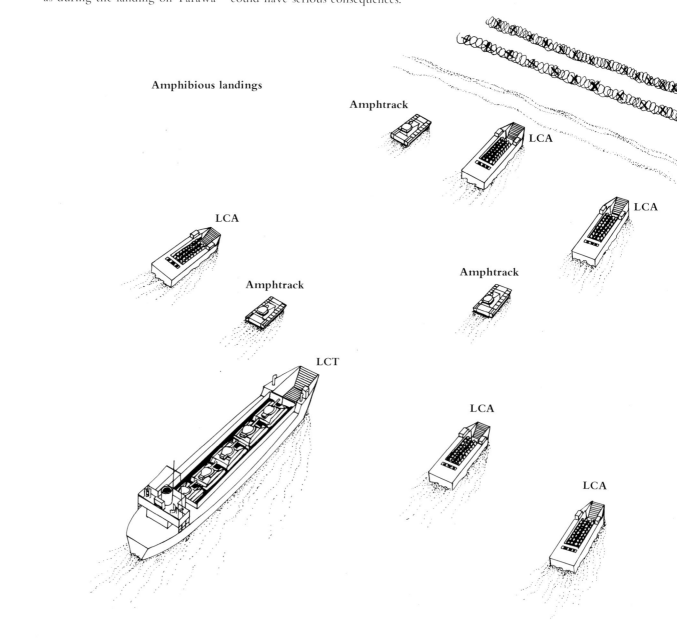

Amphibious landings

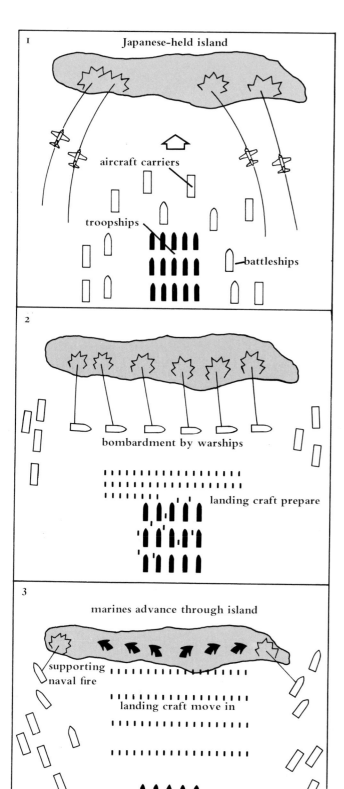

1

Japanese-held island

aircraft carriers

troopships

battleships

2

bombardment by warships

landing craft prepare

3

marines advance through island

supporting naval fire

landing craft move in

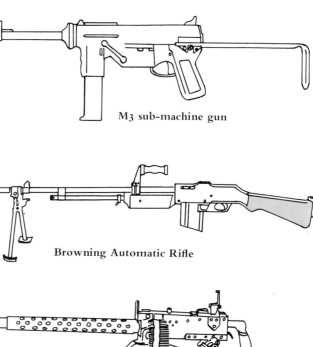

M3 sub-machine gun

Browning Automatic Rifle

M1919A4 .30 inch Browning machine gun

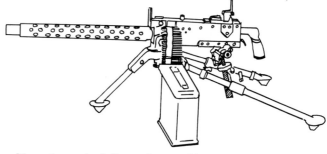

Once ashore the marines had to rely on the weapons they carried with them, and quick-firing reliable guns were essential, if only for the boost they gave to the morale of men faced with the desperately courageous Japanese. The M3 sub-machine gun was not well liked, partly because it was light and had a poor balance, but in spite of its nickname (the 'grease gun') it was very effective, being all but immune to dirt, mud and water. The BAR (Browning Automatic Rifle) was more popular, although its 20-round magazine was too small for many situations. The standard machine gun was the air-cooled .30 inch Browning, which served throughout the war as a thoroughly dependable weapon.

the beach. Unfortunately, the marines had experienced difficulty in loading the newly designed 'amphtracks' (armoured amphibious tractors,) which would take them in to shore – normal landing vehicles could not be used because of the reefs around Tarawa and the strength of the enemy defences – so the assault force was fifteen minutes late. The interval between the end of the shelling and the arrival of the infantry gave the Japanese plenty of time to climb out of their bunkers and man the 200 assorted guns on the beaches. Most of the American assault units suffered fifty per cent casualties on the beachheads, and by nightfall the marines had still not been able to link up to form a continuous front. With accurate rifle, artillery and naval gunfire and the occasional use of flamethrowers, however, they were able to reach the opposite shore on the following day, dividing the Japanese defence forces in two, and the battle was over on the afternoon of 23 November. 'The stench, the dead bodies, the twisted, torn and destroyed guns . . . are things which I shall long remember,' wrote one marine. The island had indeed been captured and the enemy garrison wiped out – the marines took only seventeen wounded Japanese prisoners – but the cost had been higher than expected: over 3,000 men killed or wounded. In the future closer support from naval and air bombardment would be needed for success in an amphibious assault.

The lessons learned at Tarawa during the first amphibious assault were soon put into practice in the assault on Kwajalein atoll in February 1944. In particular, the liaison between the infantry landing force and the naval and air support was reorganized, and the navy gunners' task was changed from neutralization of the landing area to destruction of individual defences. The support from the ships and planes was now timed to stop immediately before the men hit the beach, and not sooner; it had been found that fewer casualties were caused by accidental hits on the invading forces than by allowing the defenders time to man their guns. More and better 'amphtracks' and the use of heavy demolition charges and napalm were recommended. By the summer of 1944, the Americans were ready for attacks on the Mariana Islands of Saipan, Tinian and Guam, possession of which would put the new B-29 'Superfortress' bombers within range of Japan; the enemy was well aware of the threat and had garrisoned the three islands with more than 60,000 veteran troops.

Two marine divisions – the 2nd and the 4th – were assigned to the conquest of Saipan, together with an army division which was held in reserve. Each division was reorganized into a reinforced unit of 22,000 men, including headquarters, tank, pioneer, and engineer battalions, an artillery regiment, three infantry regiments, and service troops. Flame-throwers – both hand-held and tank-mounted – were added to the armament for use against the enemy's concrete bunkers. This time, the naval and air preparation for the assault lasted four days, and a flotilla of empty landing craft was launched against a harbour in the north of the island as a diversion

The moment of truth for the marines when they had landed was the advance off the beach to take on Japanese strongpoints inland. Although often under heavy fire, the US troops could not allow themselves to be pinned down: they had to move out. Shown here is that moment on Tarawa in 1943 (above) and on Guam in 1944 (right).

while the real invasion got under way in the south.

After three weeks' bitter fighting the marines had control of the island, resistance ceasing on 9 July 1944. Despite a resolute defence by the Japanese on Tinian, the island fell on 1 August, and was followed by American victory on Guam nine days later.

Guam was costly in American lives, nearly 2000 killed and 7000 wounded, but all three of the amphibious assaults in the Marianas had been successful; now the Central Pacific campaign was over and the bombers could begin striking at Japan's heartland while the marines moved on to the Western Pacific and their greatest battle of the war: Iwo Jima.

Iwo Jima is a tiny speck of volcanic ash and dust only eight square miles in area, lying on the route from the Marianas to Japan. It therefore provided a base for Japanese interceptors to waylay the B-29 bombers, and this threat had to be eliminated; by doing do, the Americans would also gain a useful airstrip for emergency landings. The doctrine of amphibious assault had been tested and developed in a number of campaigns, and the marines knew that it worked (although at more

cost in lives than had been anticipated); they were well prepared, with more than 450 ships and a quarter of a million men assigned to the task of capturing the island.

Iwo Jima was subjected to about two months of aerial bombardment in late 1944 and early 1945, but this made little impression; the enemy defences were well dug-in, and napalm bombs would not burn the ashes and rock of the island. D-Day was fixed for 19 February 1945, and the naval guns opened up four days before this, although the marine commander had wanted a much longer period of bombardment: 'I felt certain we would lose 15,000 men at Iwo Jima', he said. 'We were to land 60,000 assault troops, and the estimate that one in every four would be dead or wounded never left my mind.' Nevertheless, the men of the 4th and 5th Marine Divisions moved in at 9 a.m. on 19 February without meeting enemy fire, although they did have two unpleasant surprises awaiting them: the first was a 4.5m (15 foot) terrace behind the beach which the vehicles could not climb, and the second was the powdery ash of the island which made running impossible as the men sank up to their ankles. As one marine recalled after-wards, 'I arose to a crouch and tried to sprint up the terrace wall, but my feet only bogged in the sand and instead of running I crawled, trying to keep my rifle clean but failing.' As the marines inched their way along the beaches and up the terrace, the Japanese opened up with mortars and artillery from pill boxes on Mount Suribachi, the extinct volcano at the southern tip of the

island on the left of the invasion beaches, while even tougher fighting was taking place in the north of the island.

The 28th Marines, a unit of the 5th Marine Division, was given the task of capturing Mount Suribachi; it took four days before the base of the mountain was surrounded. First the marines' tanks and infantry would begin firing at an enemy pillbox while a flame-thrower team crept up to souse the entrances with flaming petrol, then the infantrymen would move in with hand grenades, and demolition teams would later blow up the remains of the fortification. On 23 February a 40-man combat patrol was asked to secure the summit of the mountain, and they took along a small American flag; when they reached the top, a small force of Japanese defenders gave them a warm reception but the flag was nevertheless tied to a piece of disused pipe and raised at 10.15 a.m. Suribachi's dominating position ensured that the dramatic gesture was seen by thousands of marines and sailors on and around the island, and the Secretary of the Navy (who had been travelling incognito with the expedition) turned to the marine commander and said: 'The raising of that flag on Suribachi means a Marine Corps for the next 500 years.'

The platoon had eliminated the Japanese force on the mountain-top, and shortly after this a marine brought up a larger and more impressive flag to replace the first one. As this was being raised – a difficult task in the loose soil – the event was photographed by Joe Rosenthal and the resulting picture was one of the most famous ever produced. Of the six marines in the photograph, half would be killed in action on Iwo Jima; the other three were called home to participate in a War Bonds campaign and a series of patriotic rallies which added significantly to the marines' image. The photograph was used on millions of posters, on War Bonds, and on postage stamps, and served as the basis for the largest bronze statue in the world, the Marine Corps War Memorial in the national cemetery at Arlington, Virginia.

The clearing of the rest of Iwo Jima occupied another three weeks, and the campaign cost the marines 6000 killed and 19,000 wounded; only 216 Japanese survived. It was the biggest and bloodiest struggle in the history of the marines, but it proved that amphibious assault could work against even the stubbornest defence – and the landing strips on Iwo would save the lives of thousands of US Air Force crewmen forced to make emergency landings during the rest of the war. As one naval commander commented, 'It is fortunate that less seasoned or less resolute troops were not committed; and Admiral Nimitz testified that 'on Iwo Island, uncommon valour was a common virtue'. Careful and sufficient pre-invasion bombardment of the target was necessary for an amphibious assault to succeed but the battle was won primarily by the marine infantry assault teams with their supporting tanks and engineers.

An even larger and bloodier campaign against the Japanese was still to come at Okinawa, and the marines would play an important part in it, but this time the army would be in charge while the bulk of the marine fighting strength rested after Iwo Jima. The Okinawa campaign could not have come about without the lessons learned by the marines – having tested and proved their doctrine, they were now able to pass on their information to the other forces. Marine airmen also played an important role in the battle for Okinawa; these fighters of the 2nd Marine Air Wing provided close air support for the advancing army and marine ground troops. By the end of the campaign on 22 June the marines had suffered more than 3000 killed and 15,000 wounded. The total number of dead and wounded marines in the Second World War was more than 86,000, and the toll would have been considerably higher if the planned invasion of Japan itself had been necessary.

During the inter-war period there had been some doubt as to what role (if any) the marines could serve in a future war and there had been talk of relegating the marines to a defensive police role. Determined to avoid this fate the marine corps enclosed the 'island-hopping' strategy of the naval planners as a means of ensuring their survival and eventual expansion; and their skills in amphibious warfare were rapidly deployed in the Pacific once America had recovered from the shock of Pearl Harbor. The string of victories that the marines gained, from Guadalcanal to Okinawa, guaranteed them a sure position in the American military establishment despite some post-war quibbling.

The marines came out of the Second World War with their fighting reputation more enhanced than ever, and the reasons are not hard to find. They fought in a new and distinctive way against a single enemy, and they did not need the cooperation of the Allies; the marines' victories were strictly *American* victories. Their campaign at Guadalcanal had been the first American offensive of the war, and marine airmen there had been the first to fight against the Japanese Zeros. The mistakes in planning at Tarawa had not affected the courage and success of the leathernecks despite what were, for America, heavy losses. Even the fact that teams of reporters and photographers served with the marines as enlisted men, in contrast to the semi-officer status they possessed in the European campaigns, helped to convince the public that they were getting first-hand information from 'up-front'. The lucky break of the Iwo Jima flag-raising picture came near the end of the war, but by this time the marines had amply demonstrated that they could indeed (in the words of their hymn):

'. . . fight our country's battles
In the air, on land, and sea'.

The most famous moment in the history of the US Marine Corps – the raising of the Stars and Stripes, 'Old Glory', on Mount Suribachi, Iwo Jima, photographed by Joe Rosenthal. The patrol was led by Lieutenant Harold G. Schrier. This photograph has since become something of a cliché, but the raising of the flag did symbolize the determination of the corps to establish itself as a major part of the American military machine.

18
THE VIET MINH

Since the end of the Second World War, guerrilla warfare has been a constant factor in military affairs, through a combination of the retreat of the western powers from their colonies, the rise of local nationalist movements, and the encouragement of such movements by the great Communist states of China and the Soviet Union. Guerrilla soldiers are self-consciously different from most of those we have examined so far. According to the theories of Mao Tse-tung, they should exist as part of the civil population, not forming a separate military society, and should be a thoroughly politicized force. Yet while eschewing many of the traditional structures of the military life, guerrilla soldiers can be very effective, even against the very best regular forces, for they embody some of the main elements necessary for any successful army. The soldiers of the Viet Minh, for example, whose striking victory over French forces in 1954 was the greatest single confirmation of the power of a guerilla army, were experienced well-trained men (and women), who knew what was expected of them and how to do it; they had confidence in their leaders and accepted completely their place in the organization of resistance to the French.

The Indochinese Communist Party was only one among many Vietnamese nationalist groups opposed to French colonial rule during the 1930s, but in 1941 the party adopted a milder attitude on the principle of land-confiscation: only land belonging to traitors (collaborators with the French) was to be given to the peasants, and 'patriotic' landowners would be allowed to keep their property. On this basis, the party invited other political groups to join in a united front for the struggle against the French, and in May 1941 the Revolutionary League for the Independence of Vietnam was formed; it was dominated by the Communists and usually referred to by the abbreviated name of Viet Minh. The guerrilla forces of the Viet Minh fought against the Japanese occupation forces in Vietnam during the Second World War, rescued American airmen who had been shot down, and supplied intelligence reports to the Allies;

A Viet Minh Guerrilla fighter. Ho Chi Minh's forces were, for the most part, indistinguishable from the civil population in which they operated.

naturally enough, they received money and weapons from the Americans at this time. The sudden collapse of the Japanese in 1945 gave the Communists a unique opportunity – if they could move fast enough: the party conference in August resolved 'to lead the masses in insurrection in order to disarm the Japanese before the arrival of the Allied forces in Indo-China; to wrest power from the Japanese and their puppet stooges and finally, as the people's power, to welcome the Allied forces'. By presenting the Allied occupation forces with a *fait accompli* – an independent national government already in control of the country – they hoped to avoid a re-imposition of French rule and achieve a bloodless Communist victory.

This was a vain hope, however. As the Allies took control of South-East Asia, the Viet Minh and the French engaged in long, but fruitless negotiations about the future of Vietnam. Eventually, on 19 December 1946 the Viet Minh blew up the Hanoi power station, cutting off all electricity in the town, and launched wave after wave of militiamen at French installations throughout Vietnam. The Indo-China War had begun.

Although some of the French garrisons were not relieved for two or three months, the French were generally able to maintain control of the cities in Vietnam after the first few weeks of the war. The countryside was another matter. With their relatively small numbers, the French could not effectively police the countryside, especially as the Viet Minh continually attacked their lines of communication, blowing up bridges, eliminating French roadblocks, and ambushing patrols. Nor could the French intern the entire Vietnamese race, and areas populated only by peaceful farmers during the day became Viet Minh strongholds every night. In the north, around Hanoi, the French at first managed to clear the guerrillas out of the Red River delta, but they were unable to reach the Viet Minh bases in the mountains to the north and east – the bases which housed the government of the Democratic Republic under Ho Chi Minh, as well as the principal training camps and political education centres of the insurgent forces. It was here that Vo Nguyen Giap and his colleagues built up the main force which would clinch victory for the Viet Minh once the time was ripe.

211

At the beginning of the war in December 1946, Ho Chi Minh had called on the entire nation to resist the French:

> Every Vietnamese, regardless of sex or age, regardless of religion, party affiliation, or nationality, must stand up and fight the French colonialists in order to save the fatherland. Let anyone who has a gun use the gun and anyone who has a sword use the sword. Those who have no swords should use picks and sticks.

In practical terms, Communist guerrilla principles (as codified in neighbouring China at this period by Mao Tse-tung) called for three types of armed forces: the militia, the regional troops, and the regulars.

The militia were, in Giap's words, 'the broad armed forces of the labouring people who are still engaged in production'. All inhabitants of a village, a city ward, or a large factory were expected to arm themselves with any available weapons to protect themselves and to engage in 'armed propaganda' under the guidance of local party officials; their duties included the distribution of Viet Minh literature, the assassination of collaborators with the French, and the collection of taxes. They were also expected to supply guides and intelligence reports for the regional and regular forces. Few of the militia were entrusted by the Viet Minh leadership with captured weapons, and they received little training in strictly military problems; political indoctrination, however, was continuous. Naturally, most of the militia duties tended to fall upon the young men – and in a village of forty persons, for example, there would be about six of them – even though all the inhabitants might be called upon to help when needed. These young and active militiamen were also the main source of recruits for the other two branches of the Viet Minh, the regional forces and the regulars.

The regional forces were composed of part-time soldiers, spending some weeks fighting and some weeks working on the land. But although they were 'part-timers', they were by no means looked down upon by the Viet Minh leadership; in fact, until the closing stages of the struggle against the French, they were the ones who did the bulk of the fighting. They were organized on a provincial basis to operate in their own areas, although they might be called up to fight outside their own province on occasion. Like the militia they were unpaid and did not wear uniform. They usually trained for two or three days a week when not taking part in their own small campaigns. A regional force would usually be of regimental strength, although the men rarely operated in units larger than platoon or company size. They operated mostly at night, since they knew their home districts well. Sometimes they would attack French outposts and watch-towers, but much of their time was spent in arranging ambushes and booby-traps, blowing up bridges, and digging trenches and immense labyrinths of tunnels in cooperation with the militia forces.

The great strength of the regional forces lay in their acceptance of their role, and their patience. They knew that their task was to harass and to harry, to wear the French down without necessarily thinking in terms of a decisive battle. The Viet Minh leaders were at pains to make this point clear: Truong Chinh, the Communist Party Secretary, wrote in 1947,

> The guiding principle of the strategy of our whole resistance must be to prolong the war.

This was because,

> if we compare our forces with those of the enemy, it is obvious that the enemy is still strong, and we are still weak. The enemy's country is an industrial one – ours is an agricultural country. The enemy has planes, tanks, warships; as for us, we have only rudimentary weapons . . . If we throw the whole of our forces into a few battles to try and decide the outcome, we shall certainly be defeated.

General Giap wrote that, through a protracted war,

> the enemy will pass slowly from the offensive to the defensive . . . Thus, the enemy will be caught in a dilemma: he has to drag out the war in order to win it and does not possess, on the other hand, the psychological and political means to fight a long drawn out war.'

Knowing that their activities would, of necessity be small-scale and would lead to victory only gradually, the regional forces were psychologically prepared for the lengthy war of attrition, which was to be the main feature of the struggle. And politically, too, with the sophisticated organization of the Communist party as their support, they were prepared to sit out reverses, and to carry through the campaign, marked by atrocities and terrorism, to extend Communist influence throughout the countryside.

Giap's 'main force' was composed of regular troops, well-trained, uniformed, and paid – at first in kind, later in cash. They were armed with captured French rifles and light machine-guns seized in the north during the Chinese occupation. At the outbreak of the war they had a minimum of 60,000 rifles and 3000 machine guns; some authorities put the figures at 100,000 and 8000 respectively. They also had plenty of ammunition, grenades and explosives. A French observer commented on the similarity of uniform among the various ranks of the Viet Minh regulars:

> There is nothing to distinguish their generals from their private soldiers except the stars they wear on their collars. Their uniform is cut out of the same wretched material, they wear the same boots, their cork helmets are identical and their colonels go on foot like their privates. They live on the rice they carry on them, on tubers they pull out of the forest earth, on the fish they catch and on the water of the mountain streams.

During the 1940s most of the fighting was done by the regional forces while the regulars were being trained in the mountain bases north and east of Hanoi. The men had normally risen from the militia and regional forces to join the regulars, and they already knew something of military discipline, weapon-handling and political theory. In the mountains they would undergo an intensive course of training in such aspects as patrolling, infiltration, and dispersion, often under Western-educated Vietnamese or Chinese instructors. They would also have the opportunity to participate in 'democracy exercises', meetings at which everyone had a right to speak and question the leaders about political, military or economic plans. In theory, any soldier could contradict his officers at these meetings, but the officers' greater education and intelligence generally earned them respect from the soldiers and the right of criticism was rarely abused. Self-criticism, on the other hand, was actively encouraged; we have already seen that General Giap was forced to admit that 'all the conceptions born of impatience and aimed at obtaining speedy victory could only be gross errors', despite the fact that he had once been their main advocate. The men were also urged to inform on their comrades when any backsliding from agreed principles was detected.

Good officers were, at first, in short supply. There were training courses at the mountain bases, both for platoon and company leaders and for the more senior grades, but many candidates were unsuccessful; officers who were criticized too often at the 'democracy exercises' were considered to have shown insufficient leadership and were removed. Those who retained their posts

Below *Women played an important part in Ho Chi Minh's forces. Their activities working to supply the fighting troops and gathering information were invaluable.* Bottom *Guerrillas on the march. The speed with which Giap was able to deploy his troops confounded all French attempts to maintain a strategic initiative.*

Tunnels and Booby Traps

The war of the Viet Minh against the French was one in which major engagements were rare, and concealment, surprise, booby traps and ambushes were the main weapons of the insurgents. Extremely complex tunnel systems were used to harbour the Viet Minh; some were miles long and could take substantial units for weeks on end. Some of the major elements are shown in the diagram below. The main entrance is underneath a hut; it has a double door as protection against grenades being thrown in. There are other entrances at different points, including one from the river, and ventilation holes could be widened to provide emergency exits. Weapons and food were stored in special chambers.

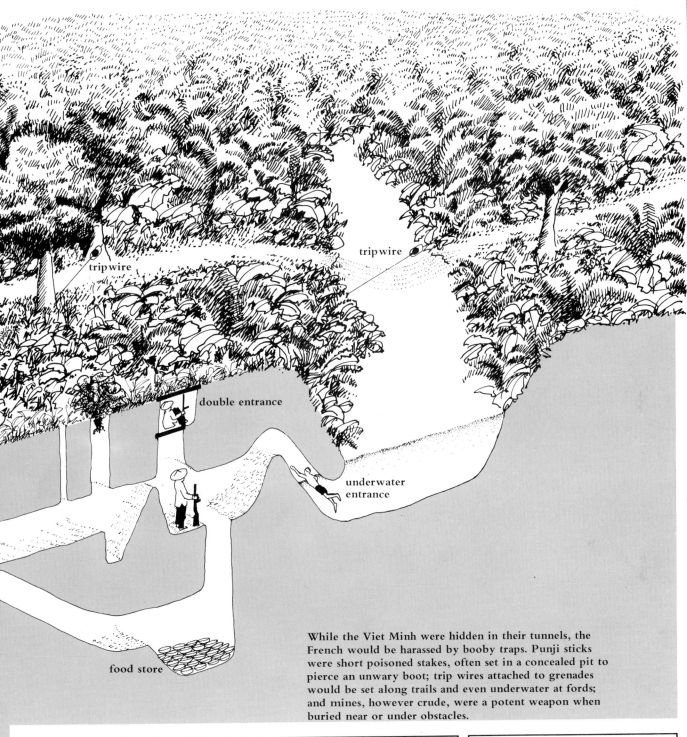

tripwire

tripwire

double entrance

underwater entrance

food store

While the Viet Minh were hidden in their tunnels, the French would be harassed by booby traps. Punji sticks were short poisoned stakes, often set in a concealed pit to pierce an unwary boot; trip wires attached to grenades would be set along trails and even underwater at fords; and mines, however crude, were a potent weapon when buried near or under obstacles.

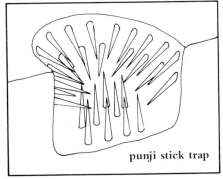

punji stick trap

tripwire attached to grenade

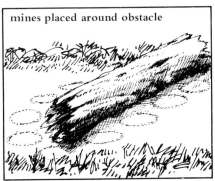

mines placed around obstacle

acquired battlefield experience by an exchange system with the regional forces. As time went on, less notice came to be taken of criticism from the ranks, and fewer officers were demoted – especially since many of them were now battle-hardened leaders.

A distinctive feature of the Viet Minh, as in other Communist-led armies, was the parallel military/political chain of command. Every military leader, right down to platoon level, had his political counterpart, and the political officer always had the last word. He was examined and trained with great care, and was expected to demonstrate his devotion to duty at all times and set an example to the men. The political officer was assisted by small cells of five or ten men in each unit in which each man was responsible for the conduct and political reliability of all the others in the cell. Political officers and cells existed in all three forces – the militia, regional troops and regulars; at each regional headquarters, for example, there was a political office which reported directly to the leadership of the Viet Minh without going through the military chain of command. This system naturally allowed Ho Chi Minh and his colleagues to make frequent re-assessments of the military and political reliability of their forces at all levels.

The three types of Viet Minh troops were paralleled by the three stages of revolutionary warfare as enunciated by General Giap: defensive, equilibrium and offensive. The Viet Minh attack in December 1946 was merely a short prelude to the real struggle, which began with the French counter-attack. Giap realized that the French would rapidly gain control of the cities and towns, and there was little to be gained by contesting them. As Mao had put it, in the first stage of a revolutionary war 'when the enemy advances, we retreat'. By February 1947 the Viet Minh had fled to their bases in the mountains to prepare for the second stage of the war: 'The enemy camps, we harass.' This was the particular responsibility of the militia and regional forces. Viet Minh political agents would be sent out to the villages, especially behind the French lines, to organize the inhabitants, supervise the construction of secret supply dumps and tunnels, and to identify the people with the struggle of the Viet Minh. Echoing Mao's precept that 'the people are the water and our army the fish', Viet Minh regulars would discard their uniforms and weapons whenever there was a prospect of encirclement by the French; in a few seconds a soldier could transform himself into a peasant and escape detection, so long as he had the support of the local people. As local guerrilla operations were stepped up, the French found it more and more difficult to find the time or resources to administer areas of the countryside; they discovered that they had control over only the land on which they were standing. As Giap wrote:

> Our guerrillas and government take over within gunshot of their strongpoints. In most cases, we are actually besieging them. The minute a soldier pops out of a blockhouse, he faces guerrilla fire.

This campaign of attrition was in full swing in 1948 and 1949. Meanwhile, events outside Vietnam were influencing the political composition of the Viet Minh: Communist uprisings in Malaya and Burma had broken out early in 1948, and in December of the following year the Chinese Communists under Mao reached the Vietnamese border by defeating the Chinese Nationalists. With this encouragement, in 1949 the Vietnamese Communists took complete control of the Viet Minh army and the government of the Democratic Republic, throwing out the leaders of the non-Communist nationalist groups which had worked with them since 1941. In return, Mao's forces sent thousands of extra rifles, mortars and machine guns, together with tools and machinery, and new training camps were set up in southern China. By 1950 the Viet Minh were ready to launch a series of strong local attacks on the French forts along the Chinese border, using forty battalions of regulars armed with mortars and artillery. Within a few months they had succeeded in taking all the forts and had captured 8000 rifles and 1000 machine guns – enough to equip another Viet Minh division. As they moved down into the Red River delta around Hanoi, however, the Viet Minh were unable to break through prepared French positions, and were driven back into the mountains. It was this defeat that taught Giap his lesson, and he would not send his regulars out into the plains to fight a conventional battle against the French again until the end of the war.

For the next three years, the Viet Minh reverted to local guerrilla warfare and concentrated on infiltrating areas behind the French lines. 'Struggle, fail, struggle again, fail again, struggle again till victory . . . that is the logic of the people,' Mao had advised his Chinese followers. Even if ninety per cent of the revolutionary army and its bases and territory were wiped out, it would only be a partial and temporary setback, according to this theory of warfare. The result of the guerrilla struggle of attrition, therefore, was inevitable: losses on both sides

The plain of Dien Bien Phu, where the French thought they could make their material superiority tell against the guerrillas of the Viet Minh.

mounted, but the Viet Minh's could be replaced while those of the French could not. Giap summed up the French dilemma in 1954:

> Either they try to extend their strongpoints once again, with their depleted manpower, in which case they must spread themselves thin and lay themselves open to new attacks that we can launch in regimental or division strength, or else they can try to reduce these strongpoints and consolidate them, which frees territory and population to us.

The French had by now committed nearly 190,000 men to the struggle for Vietnam – a quarter of them French and the rest Legionnaires, North Africans, and loyalist Vietnamese – but fifty per cent of this total were tied down in the defence of strongpoints and could not move out to 'take the war to the Viet Minh'. By 1953 the Viet Minh, on the other hand, comprised 125,000 mobile regular troops, more than 75,000 men in the regional forces, and between 200,000 and 350,000 supporting militia.

The armistice which ended the fighting in Korea in July 1953 freed thousands of tons of Communist Chinese equipment which was immediately sent south to the Viet Minh. Now, with the balance firmly in his favour, Giap was able to bring his regulars out of the hills once again. Between December 1953 and January 1954 the Viet Minh invaded Laos and threatened to cut Indo-China in two; in February they rapidly infiltrated an entire division of three regiments inside the French lines in the Red River delta. The French responded to the invasion of Laos by moving a strong concentration of troops westward from the delta towards the Laotian border in a deliberate attempt to draw the Viet Minh regulars into a pitched battle near the base of Dien Bien Phu. This time,

Giap's regulars were ready and willing to confront them. First the French supply lines were attacked by the division which had infiltrated the delta. The men crept through the sewer system of the French military airport, which was heavily defended against ground attacks because of its importance as a supply base for the Dien Bien Phu expedition, and destroyed thirty-eight aircraft. In March the road link between the fortress and the capital was attacked every day, and meanwhile the Viet Minh prepared to besiege the French in Dien Bien Phu itself. 50,000 men were employed in bringing a total of 200 medium field guns up to a captured hill overlooking the base – mostly by dismantling the guns and moving the parts by bicycle. Tunnels were dug into the hillside from the rear and the guns pushed through so that only their muzzles were exposed; these could hardly be seen, let alone hit, by the French defenders. Their own guns, sited out in the open to give good fields of fire, were smashed one by one, and the monsoon rains in April helped reduce the French trenches and foxholes to mud. As the garrison was slowly starved of supplies, the Viet Minh troops brought up rocket launchers and began tunnelling underneath the French positions, as well as launching 'human wave' attacks against the fortress. The last French gun at Dien Bien Phu was silenced on 8 May, and by late July the war was over – permanently for the French, temporarily for the Viet Minh.

The Viet Minh succeeded in expelling the French from Vietnam partly because the French had no coherent political programme of their own to present to the Vietnamese people and had little support among them; the Viet Minh, on the other hand, presented themselves as nationalist liberators and laid great emphasis on 'correct' behaviour in dealing with the people; soldiers were under strict orders to be fair, polite and scrupulously honest in their dealings with civilians. Another factor in favour of the Viet Minh was the nature of the campaign – a guerrilla resistance to an occupying power whose base was thousands of miles away; in such a struggle, the advantage must lie with the insurgents. Lastly, the Viet Minh learned (partly by trial and error) when to use guerrilla tactics and when to use regular troops. The abortive campaign in the delta in 1950 taught them that guerrilla fighting must be used to wear down the enemy's military strength and will-power until the advantage is clearly on the insurgents' side, at which time the regulars must be used (as at Dien Bien Phu) for the final thrust to clinch the victory. Guerrilla warfare is by no means a universal recipe for military success, and if the conditions are not right – for example, if the insurgents fail to gain the support of the majority of the people – then it can be put down, as the British succeeded in doing in nearby Malaya. But the lessons learned in the struggle of the Viet Minh against the French would provide a pattern for the eventual Communist success throughout Vietnam in the 1970s, and many of those lessons and tactics would also be adopted by revolutionary movements throughout the world after the mid-fifties.

19
THE FRENCH FOREIGN LEGION

The French Foreign Legion is the most famous mercenary force in the world. In many ways, however, it is untypical of mercenary forces, for its loyalty, devotion to duty and unflinching service in the interests of a single country – France – are rarely found in mercenary bodies. The Legion's greatest test has been in the two guerrilla wars of the post-war period in Indo-China and Algeria, which together lasted from 1946 to 1961. In both these campaigns the Legion acquitted itself superbly, in spite of the fact that the causes it was fighting for were defeated.

In this new form of warfare, where a whole population was a potential enemy, and the struggle seemed an infinite toil with no possibility of a decisive victory, the regular soldier was at a considerable psychological disadvantage; his training had not prepared him for such a task. But the Legion fought on in conditions where other forces cracked and became unreliable.

The strength of the Legion has always lain in its abilities to turn recruits from many lands into devoted members of a military society, which rules their whole lives and for which they are willing to perform often reckless acts of bravery. It is not patriotism, revolutionary zeal or mere discipline which inspires the legionnaires, (although discipline certainly has its place), but pride in the tradition of professional soldiering and corporate solidarity which the Legion embodies.

The recruitment procedure of the Legion has set it apart from the units of almost all other nations. Even in the post-war years a man could join up for an initial period of five years, without any identification papers being demanded; any name he cared to give would be accepted. A rigorous medical examination would establish the recruit's fitness: anyone who might appear to be between the ages of eighteen and forty (for which again no documentary proof was demanded) could be a Legionnaire.

The legionnaires who fought in Indo-China and Algeria came from many lands – there were reputedly over fifty nationalities serving during this period. But although it is impossible to get exact figures, it seems

very likely that the largest contingent was German. The Legion accepted many recruits after the Second World War who were given only the most cursory screening, and recruiting offices were set up in the French zone of occupied Germany. It has been calculated that about eighty per cent of the NCOs were German. Many Italians, too, were welcomed with open arms.

After all European wars, there has been an influx of the defeated armies' troops into the Legion – White Russians during the early 1920s for instance. What was ironic about the influx of Germans in 1946 was that many of these men had participated in the greatest military humiliation France has ever suffered (in June 1940) and now presented themselves for a long period of service for that same country. Many of these men also probably committed atrocities against the population of occupied France, and were joining up in order to ensure their own anonymity and escape from a war tribunal. Such individuals came from all over Europe: there is the well-documented story, for example, of the Romanian fascist Stanescu who had participated in the massacre of Romanian Jews, and was pursued by a young Israeli who had survived this slaughter. Stanescu, having changed his name, was a corporal in the Legion in Indo-China when

Left A legionnaire in the combat uniform worn in Algeria and a 'saharienne'.
Right The traditional Legion: in Morocco in the 1920s.

the Israeli, having himself joined the Legion, finally caught up with him and killed him. Frenchmen accused of collaborating with the Germans in atrocities against their countrymen are also known to have joined up just after the war.

Men who joined the Legion were generally very willing to fight: and those who joined in the late 1940s often had enormous experience of modern warfare – having served with the Axis armies in Western Europe, North Africa, or against the Soviet Union, that most ferocious of campaigns. Such men were excellent raw material for a force such as the Legion. Many preferred to carry on as part of an army rather than scrape a living in a defeated country; some may just have enjoyed the institutionalized life of a soldier. But all, without fail, were thoroughly indoctrinated in the spirit of the Legion, which turned them from men who merely wanted a military career into men who would die for their unit.

The Legion has never tried to make its recruits feel any loyalty towards their ultimate employer, France. The Legion itself, its history and its traditions, are made the focal point. The new men were told repeatedly that the Legion was unquestionably the finest force in the world, and that it was up to every individual to reinforce and add to its fighting tradition. A major part of this process of indoctrination was embodied in the Legion's ceremonial. The original regimental colour is still in existence, and each regiment had its own colour, most of which were heavily decorated. The cavalry and Saharienne companies had the smaller fanions. Different coloured fourragères (lanyards) were worn on the shoulder as marks of distinction: the 3rd REI (Régiment Etrangère d'Infanterie) wore dark red for its achievements in the First World War, for example.

The history of the Legion has always been an important part of the life of the troops. The highlight of the Legion year was 30 April, the anniversary of the fight at Camerone in Mexico, where in 1863 as part of a French expeditionary force fighting a forlorn war to impose the emperor Maximilian on an unwilling population, Captain Danjou and forty-six legionnaires fought all day against over a thousand Mexicans, until they were finally overwhelmed. On the evening of 29 April, there would be a torchlight parade at the Legion's base at Sidi-bel-Abbès, and on the following day, legionnaires everywhere were paraded and read out the official account of the battle of Camerone. At Sidi-bel-Abbès, the false hand of Captain Danjou, found in the ruins of the farmhouse, was borne in front of the men.

The Legion has, too, a distinctive style of uniform. By 1914, the typical North African wear of blue jacket, red trousers, blue greatcoat and off-white trousers had become standardized, and although this was naturally not worn in Indo-China or Algeria, the ordinary legionnaire's ceremonial uniform was still unique: the white képi, blue cummerbund, epaulettes, white gloves, green tie and anklets. And there were always the more exotic elements – the pioneers with their beards and axes,

and the Saharienne companies, in which sandals, black trousers, white tunics and blue cloaks were worn.

Of course, such official ceremonial does not necessarily make a profound impression on the individuals subjected to it. But in the case of the Legion, it was reflected in semi-official and unofficial traditions which go deep into the lives of the men. The songs, for instance, are quite unique. The most famous is the Boudin (Black Pudding); the chorus is about the food itself. The tune is a slow march, and few armies can ever have marched solemnly at ninety paces per minute singing about sausages. At Christmas, the Legion has always made special celebration, partly no doubt because of the strong German influence (Christmas was never a great day in the French army as a whole) and a tradition of food and gifts grew up. Then there is the famous rallying cry 'A moi la Légion' (literally 'To me the Legion'), used whenever a single man or a small group were in danger – which, no doubt, was often the case in the dubious areas frequented by the men when off duty.

Underlying these official and semi-official means by which the Legionnaire was made to feel part of a unit rather than a passing mercenary, was the accent on toughness and brutality. This was the lowest common denominator which most recruits could understand, and they had to be made to feel that the Legion was the toughest fighting force anywhere. Legion training was extraordinarily hard, and Legion attitudes to the enemy could be pitiless. Accounts of the tortures and massacres perpetrated by legionnaires in Vietnam and Algeria leave no doubt about that. If it gloried in, and took to its ultimate point, that disciplined solidarity which is the basis of a combat unit, it also took to extremes an acceptance of death and brutality as the essential handmaidens of military success.

Personifying so many of the elements which create a rugged fighting unit, the Legion also suffered greatly from desertion, that most basic of military crimes. Desertion had always been a problem, from the very earliest campaigns. A force of such polyglot composition was bound to suffer in this way, offering as it did a haven to many who, little more than criminals, could not endure the rigidity of military life, and with a romantic aura attracting others who would find military discipline unbearable. Even in the bitter desert fighting in North Africa as France established her empire, there had been many instances of men going over to the other side. During the campaign against the Riffs in the 1920s, a German legionnaire deserter called Klems was one of the main military advisers to Abd'El Krim. During the two World Wars, desertion was much reduced, because of the nature of the conflicts, but in Indo-China, it was again frequent. The Viet Minh were careful to treat deserters (and often captured legionnaires) as well as possible, and made sure that all French units were aware

The 'honour and fidelity' of the Legion were severely strained during the Algerian war, but this mercenary force stayed loyal during the crises of the early 1960s.

The Legion in Indo-China

In Indo-China from 1946 to 1954, the Legion faced a great test of its ability, but it came out of the fighting with its reputation as a fighting force enhanced.

The main weapons used by legionnaires fighting in jungle, paddy field and mountain were the MAS 49 rifle, which came into service in the late 1940s, and the MAT 49 sub-machine gun (a small, light weapon with a high rate of fire).

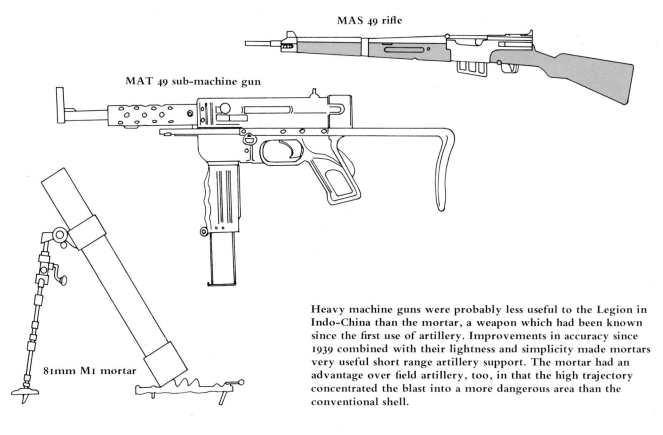

MAS 49 rifle

MAT 49 sub-machine gun

81mm M1 mortar

Heavy machine guns were probably less useful to the Legion in Indo-China than the mortar, a weapon which had been known since the first use of artillery. Improvements in accuracy since 1939 combined with their lightness and simplicity made mortars very useful short range artillery support. The mortar had an advantage over field artillery, too, in that the high trajectory concentrated the blast into a more dangerous area than the conventional shell.

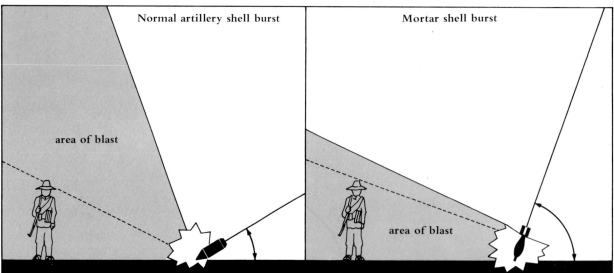

Normal artillery shell burst

Mortar shell burst

area of blast

area of blast

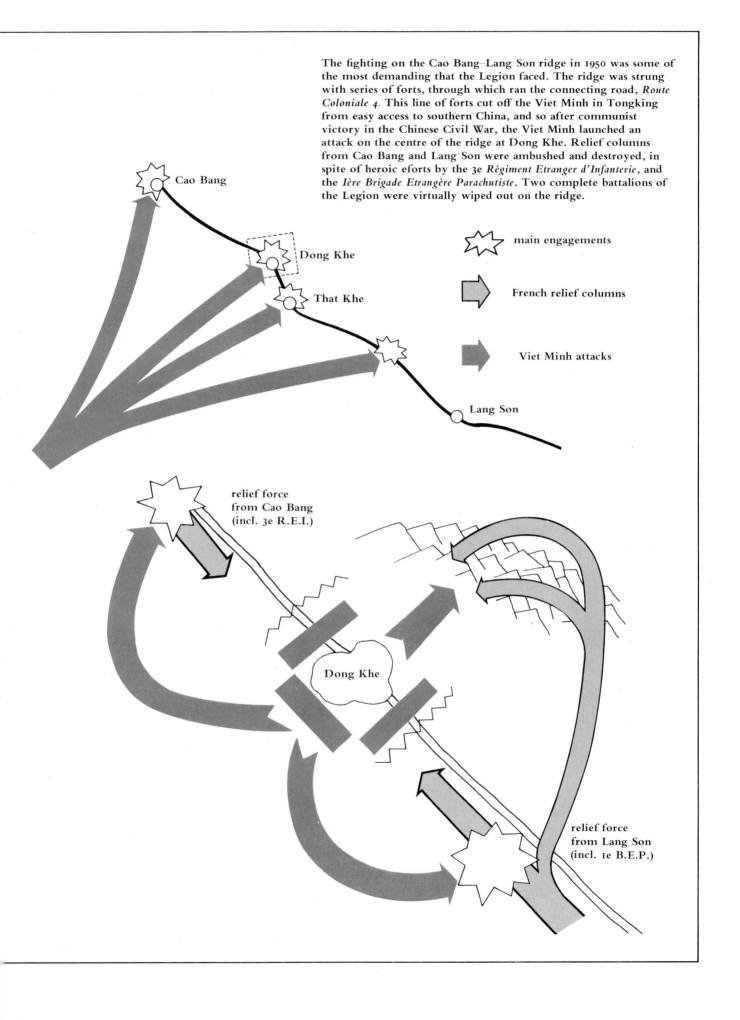

The fighting on the Cao Bang–Lang Son ridge in 1950 was some of the most demanding that the Legion faced. The ridge was strung with series of forts, through which ran the connecting road, *Route Coloniale 4*. This line of forts cut off the Viet Minh in Tongking from easy access to southern China, and so after communist victory in the Chinese Civil War, the Viet Minh launched an attack on the centre of the ridge at Dong Khe. Relief columns from Cao Bang and Lang Son were ambushed and destroyed, in spite of heroic eforts by the 3e *Régiment Etranger d'Infanterie*, and the *Ière Brigade Etrangère Parachutiste*. Two complete battalions of the Legion were virtually wiped out on the ridge.

Cao Bang

Dong Khe

That Khe

Lang Son

main engagements

French relief columns

Viet Minh attacks

relief force
from Cao Bang
(incl. 3e R.E.I.)

Dong Khe

relief force
from Lang Son
(incl. 1e B.E.P.)

of this. Precise figures are difficult to find, but some estimates put the number of deserters in Indo-China at nearly 4000. A proportion of these, probably less than 500, took up arms against their former comrades; one Austrian adopted a Vietnamese name and helped the Viet Minh as a military advisor.

In Algeria, where escape routes back to a European mother country were considerably shorter than from Indo-China, there was a steady outflow of men, encouraged by the guerrilla forces of the FLN, who followed the Viet Minh lead of treating such men well and organizing their return to Europe. Legion units had to be moved around constantly while in Algeria, and the steady stream sometimes threatened to become a flood. There was, too, especially towards the end, a spate of desertions towards the OAS, and related right wing organizations which were fighting to stop de Gaulle's attempts to negotiate a French withdrawal from the country. The Legion was as affected as any other unit would have been by the probable loss of its only home – Algeria. But what is so interesting about the Legion is that in spite of these desertions, and in spite of the terrific strain it was under, it carried on fighting obediently to the very end in both Indo-China and Algeria.

There had been a Legion regiment in Vietnam during the Second World War, but it had suffered greatly during the Japanese occupation, and was dissolved in May 1946. In February of that year, a newly created regiment, the *2ᵉ Régiment Etrangère d'Infanterie* (2nd REI)

The Legion went to Indo-China with a first-class reputation after its exploits during the Second World War, and this US Stuart tank with its legionnaire crew in Tongking in 1946 probably had fought at Alamein.
Right A patrol on a blockhouse in North Africa during the early 1950s.

had arrived, to be followed by the 13th Demi-Brigade in March. In all, seven major Legion units were to be involved against the Viet Minh.

The Legion had had experience of pacifying subject populations in North Africa during the nineteenth and early twentieth centuries, but it had never been involved in a guerrilla war such as this one, where the sophistication of the Viet Minh organization meant that every Vietnamese was a potential enemy. It was even rumoured that the most beautiful prostitutes had been purposely infected with venereal disease. The terrain over which the Viet Minh chose to do its fighting was ideal for ambushes, and for the assembly and dispersal of large bodies of infantry, hidden from the air. To maintain the logistical support necessary for its operations, the French army had to establish a network of forts along lines of communication; all of these were liable to be attacked at any time. The *Route Coloniale 4*, for example, running through Tonkin, was a prime target for the Viet Minh, especially after Mao's takeover of China, when to force the French from this road would secure communications with a friendly Communist power.

224

In July 1948 one of the great battles in the history of the Legion was fought along this road at the fort of Phu Tong Hoa. 104 legionnaires with two light machine guns fought off an assault by many times their number of Viet Minh supported by heavy mortars. Three of the four corner bastions fell, and all the officers were killed; but the NCOs led a courageous resistance, recaptured the fort, and organized a guard of honour with red epaulettes and white *képis* to greet the relief column.

In September and October 1950, the struggle for RC4 also witnessed one of the Legion's great disasters. In mid-September, eighty-five legionnaires died and 140 were captured when the fort of Dong Khe was taken, and the French high command decided to pull back from the RC4. But in the fighting to relieve Dong Khe and then to achieve an orderly withdrawal, the *1er Bataillon Etrangère Parachutiste* (1st BEP) suffered casualties of ninety per cent, and was disbanded as a unit.

In common with other formations, the Legion suffered a definite drop in morale. Desertion increased; suicide became more common, and macabre games (of men charging at each other to clash skulls, or shooting pistols at each other in darkened rooms) made their appearance. But the Legion still retained its spirit. This was partly due to the fact that in a war of small detachments, it had a vastly experienced core of NCOs, mostly German and often with ten years of combat under their belts. We have seen how at Phu Tong Hoa, the NCOs had carried on when all the officers were killed. Their status in the Legion was greater than in most other units; they were saluted by the men, and they had greater authority – such as the power to order certain punishments which in other French units could only be imposed by officers. Then too, the Legion's nature as a complete military society gave its members resources to fall back on. The legionnaires, existing in forts which they had often built themselves along roads they had constructed, managed to keep morale intact.

By the early 1950s, there were well over 50,000 legionnaires in Indo-China, and naturally the quality of the various units varied. But the basic will to fight was always present – even during the decisive defeat of the French forces at Dien Bien Phu. Over half of the garrison of 14,000 men in this ill-fated set of forts was from the Legion: seven out of twelve battalions. The strategic weaknesses of the position are now clear and the legionnaires were doomed, but they fought desperately. As the Viet Minh artillery gradually dominated the battlefield, and a form of trench warfare emerged, the mercenaries contested every inch of ground against massed attacks, even when their position was obviously hopeless. 700 volunteers from other Legion units made a parachute drop for the first time in their lives to reinforce the garrison. By the middle of April 1954, the garrison was down to 3500 men, of whom 2500 were legionnaires. On 30 April, they celebrated Camerone day, and Colonel Lemeunier (who had himself volunteered to parachute into Dien Bien Phu when the senior Legion

225

officer there was killed) read the story of Camerone over the camp tannoy. Finally, when the camp surrendered on 8 May, the 600 legionnaires in the most southerly fort *Isabelle* tried to fight their way out, although only a handful got through.

In all, in Indo-China, 314 officers and 10,168 Legionnaires died, while three times as many were wounded. Then, in November 1954, a series of attacks by the FLN forces in Algeria opened up another guerrilla war which France was destined to lose, and in which, once again, the Legion would have to bear some of the toughest fighting. As the war in Algeria opened, the Legion was being reduced in size after the disaster in Indo-China. It had been decided to scale it down to 12,000 men; recruits were given stiffer tests and re-engagement was discouraged. Men were needed to fight the new enemies, however, and so the legionnaires were put into action.

A combination of a short period of official reduction, the aftermath of the defeat in Indo-China and the start of a new debilitating guerrilla war led to a sharp drop in recruits and re-engagement. In 1955, only 2981 new men were accepted (in 1953 there had been 6327) and between September 1955 and July 1956, no man of the 13th Demi-Brigade signed up for more service. The Legion was undergoing a crisis. Men coming back from Indo-China were noticeably scarred by the experience, and the conditions of service in Algeria were unpopular. The sudden influx of troops into the country strained military

accommodation: in 1955, a barracks for 600 men might be housing 3000. The vast wastes of the Algerian interior could hide any number of guerrillas, and futile sweeping operations or long drawn out months of isolated garrison duty always left a long list of casualties, with no obvious result.

The French did have some successes: in 1957, for example, FLN terror in Algiers was quelled by French parachutists, with the Legion's 1st REP taking a prominent part in this ruthless and brutal (though very effective) operation. Then from 1958 when General Challe took over, the French forces adopted a new policy. A system of electrified fences cut the frontiers with Morocco and Tunisia (which the FLN had crossed

with impunity) and while the mass of the French conscript army garrisoned various areas, front line units – basically the Legion and parachute regiments – were used to clear rebel-held areas one by one.

This fighting was much better for the morale of the Legion, and the men could feel that they were achieving something, even though the long term gains were slight. FLN activity was damped down, but it could not be ended. Legion casualties probably reached about 1500 during each year of the war. But the fact that the new operations seemed more successful was important, and morale rose from the low point of the mid-1950s. Units found more permanent headquarters (the 13th Demi-Brigade near Bougie, for example) which fostered the corporate spirit. Techniques for fighting using airborne sweeps and making maximum use of superior fire power were developed: and the bitter legacy of the Vietnamese war, of men who had been part of a defeated army, was gradually dissipated.

By 1960, the French government had determined to give Algeria independence. This move was bitterly opposed by most Algerians of French descent – the *pieds noirs* – and by many elements in the army. It was naturally repugnant to the Legion. Its physical home had always been Algeria, at Sidi-bel-Abbès; its spiritual centre also lay in this land which it had conquered, held, and was now about to lose. In 1960, during the 'Day of the Barricades' the Legion parachutists were used to subdue rioting by the French *colons* in Algiers; but they had gone about the task with divided loyalties. All through the later months of 1960 there was unrest and desertion from the 1st REP. In April 1961, the regiment took a prominent part in the rising of the generals, an attempted *coup* led by Challe. With the 1st REP in the van, Algiers was taken over by the rebels, but the bulk of the Legion units remained loyal to the government. At Sidi-bel-Abbès, Colonel Brothier, in command of the 1st REI gave the classic professional reply to requests to join the rising: 'The Legion is foreign by definition and will not intervene in a purely French quarrel.' Algeria was irretrievably lost.

On 24 October 1962 the final 700 legionnaires who remained at Sidi-bel-Abbès paraded for the last time. All the moveable relics of the force, however, had been moved to the new headquarters at Aubagne, near Marseilles. The sword of a Danish Prince who had served with the Legion; the medals of three Hungarian adjutants who had joined up together and died within a week of each other in Algeria; the battle flags decorated with more battle honours than any other French units, and, of course, the artificial hand of Captain Danjou from Camerone; they were all removed to recreate the *espirit de corps* which had kept the Legion, this foreign mercenary force, loyally fighting on through the merciless campaigns of Indo-China and Algeria.

A Legion patrol in familiar territory: the Atlas mountains in which some of the bitterest fighting in the Legion's long history had taken place.

20
THE ISRAELI ARMY

The Israeli Defence Force and the soldiers it produces are unlike any others in the world, and this is hardly surprising. The State of Israel has, until recently, been surrounded on its three land frontiers by nations which publicly and repeatedly vowed to destroy it. Until the decisive victories of the Six Day War in 1967 the Jordanian Army held territory only sixteen kilometres (ten miles) from the Mediterranean, squeezing Israel into a tiny strip of coastal plain at its narrowest point, while Egypt had an army poised in the Gaza enclave in the south and Syria commanded the Golan Heights overlooking the country's most fertile agricultural and fishing area in Galilee. Almost all of Israel was within range of Arab long-range field artillery as well as bombers and missiles. The Mediterranean coast and the Straits of Tiran leading to the port of Eilat in the south were susceptible to blockade. The total Jewish population of Israel in 1967 numbered about two-and-a-quarter million, while the surrounding hostile Arab countries totalled forty-four million. It is not difficult to understand why Israel has a siege mentality, and why a large percentage of the national budget is still devoted to defence.

The effect of this was to reinforce the tradition of mutual support in adversity which characterized the Jews during the centuries of dispersion. Although the Israeli Army is organized along the lines of most western armies, it has a much closer involvement in civilian life than in other such forces, for in Israel it is obvious that defence is everyone's business. Even now all Israelis are regarded as being one family fighting for survival, and the outbreak of war has always prompted spontaneous acts of generosity to the armed forces from civilians, ranging from free showers for the troops to large interest-free loans or outright gifts of money for the war effort. The number of civilian volunteers for support duties has also been greater than the number of unfilled positions on each occasion when Israel has gone to war.

Left An Israeli soldier in 1967, with FN rifle and slouch hat.
Right 2 June 1967: soldiers from a Kibbutz move up to the front near Jerusalem. At this time, Israel was very vulnerable in this sector.

At the time of the Six Day War, many Israeli Jews shared the Zionist attitude to the defence of what they considered to be a God-given state. As a result there was no provision for conscientious objection to conscription. Jews (but not Arab citizens) were drafted at the age of eighteen, most women as well as all men, and served for twenty-six months in the regular army. Basic training for men lasted eight weeks and included special emphasis on night combat, quick movement, and 'roughing-it' during route marches. Later they would undergo further training in the type of unit chosen on the day of induction, of which the toughest is the Parachute Corps, the closest equivalent of a commando force, and which was the first unit to enter the Old City of Jerusalem during the 1967 war.

Although Israeli women soldiers carry weapons and are trained to use them for self-defence, they are not in fact assigned to combat duties. Instead, they fulfil service roles such as clerical and administrative work, education, welfare, medical services, translation, and police duties, freeing men for the actual fighting. They live in separate barracks, which are out of bounds to the men at all times, and it is generally felt that their presence helps restrain anti-social behaviour among male soldiers. Apart from

this, Israel has no tradition of 'men only' clubs or activities, and mixed service in the armed forces is a natural development from co-educational schools and attitudes favouring sexual equality at work and in public life.

The conscription of all Jews at the age of eighteen produces more than enough soldiers in peacetime, but nowhere near the number required in war; Israel therefore relies on a highly developed reserve system to mobilize the nation when necessary. In most countries, reservists are seen as civilians who return to the colours for refresher courses for a short period each year, but General Yigael Yadin has described Israeli reservists as 'regular soldiers who happen to be on leave for eleven months out of twelve'. They are liable for thirty-one consecutive days' service, together with one day per month or three days every three months at their commander's discretion, although this is reduced to a total of two weeks at the age of thirty-nine. Women are liable for reserve duties until they marry or reach the age of twenty-nine. All reservists return to their original units upon mobilization, and many conscripts therefore receive training in technical skills which are not needed in peacetime but would be urgently required in the event of war. During the eleven months per year when the reserve unit is 'off duty', a small permanent staff keeps track of members' addresses so that the troops can be mobilized quickly if necessary. Practice mobilizations are frequent, and can be either public (through announcements on radio and in the press) or secret (through the broadcasting of code words). Since army units are formed on a geographical basis, reservists can be called to the colours with remarkable speed; in an emergency, a quarter of a million men and women can be mobilized within twenty-four to seventy-two hours.

Another distinctive feature of the Israeli army is, of course, the part played by religion. Although the people of Israel are united by an ideal and a common racial and religious heritage, they are nevertheless very diverse in terms of country of origin, language, religious tradition and social customs, ranging from secularized Westerners to the ultra-orthodox *Hassidim* and the educationally and socially backward Yemeni Jews. Nor is orthodox Judaism a private and interior matter; it demands certain styles of dress, public observances, fasts and strict adherence to rather cumbersome dietary laws. To prevent embarrassment and tension, orthodox leaders in the early days of the Israeli Army asked for separate units for observing Jews. This policy, however, was not adopted; instead, religious precepts were written into the army regulations even at the risk of wasting the time of the non-religious. Food is strictly kosher and the sabbath is kept as a day of rest, with no training, work or sports. Soldiers may attend the camp synagogue for prayers three times a day, and for an hour and a half on the sabbath. During religious fasts they are not required to get their hair cut, and they may wear beards at all times. Every soldier is issued with a Bible when joining the army, and he may also receive articles for worship if desired. The Passover is celebrated with appropriate ceremony, and on Yom Kippur everything is closed, including the kitchens. The non-religious are not required to attend prayers or observe fasts, but they eat kosher food like the orthodox and are expected to accept the restrictions without complaint. On the other hand, as one officer has stated, 'Officially we can't eat non-kosher food, but this army also does a lot of unofficial eating.' And despite the sabbath precepts, Israeli troops will go into combat if necessary on that day as well as any other.

Partly because of the tradition of mutual dependence, partly because of the newness of the state, and partly because of Israel's economic and political organization, class barriers are not overwhelmingly evident in Israeli society. Standards of living have risen rapidly for all since 1948, and there are no landed estates or large concentrations of private wealth to be inherited. Naturally, this has its effect on relations within the army; there is, for example, no tradition of the 'officer and gentleman' type, and soldiers are remarkably free in discussion and complaint with their superiors. Most officers rise from the ranks and are not much older than the men serving under them: company commanders have an average age of twenty-two to twenty-three and battalion leaders are aged about twenty-six to twenty-eight. The authoritarianism symbolized by the sergeant-major in many other armies is unknown in Israel. Nevertheless, things have changed somewhat since the revolutionary days of the underground army fighting for independence, when many commanders made a virtue of slovenly appearance as a rejection of foreign values and the relationships of the guerrilla band were carried over into the army. In recent years there has been more stress on formality, and officers now receive smarter uniforms and eat in their own messes, but there are still relatively few external signs of authority. Saluting is rare, and an early attempt to get soldiers to say, 'Yes, Commander' when acknowledging orders was soon dropped.

One exception to the universality of conscription and the principle of 'rising from the ranks' is the Academic Reserve, in which officer candidates are allowed to complete their studies at university. The candidates join the army with the others at the age of eighteen and undergo a short basic training, after which they can return to their studies as 'soldiers on special leave' while remaining subject to military discipline. During the summer they are back with the army for officer training along with the candidates from the ranks. In the early years of the Academic Reserve system it was feared that the officers produced in this way might not fit in with the rest of the army, but they seem to have performed as well as those with less formal education who have served in the army continuously from the age of eighteen.

Left above *A column of half-tracks pushes forwards.*
Left *The great triumph of June 1967 reflected in the joy of Brigadier Rabi Shlomo Goren, Chief Army Chaplain, at the Wailing Wall.*

Aside from the qualities which any army expects of its officers, such as giving clear orders, showing initiative and maintaining the unity of his group, Israeli military leaders are expected to demonstrate talents of 'leadership' as opposed to 'command'. Officers are praised for their ability to encourage mutual help, for 'roughing it', for being good mixers, for studying the views of their men and for encouraging initiative in facing unforeseen problems. Those who rely on patriotic jargon or the attitude of 'you're in the army now' are not well thought of. Officers are generally known by their first names or even by nicknames.

The most important reason for the close links between officers and men in the Israeli army, however, is the principle of 'leading from the front' or 'follow me'. This is both a legacy of the days before 1948 when Israel's defenders were guerrillas, for whom this style of leadership is essential, and the logical outcome of Israel's dedication to the sharp and deep thrust into enemy territory. 'Follow me' should not be taken too literally, for the Israeli officer does not actually overtake his men in a headlong charge, but he does participate directly in

Right *An Israeli soldier rests on his anti-aircraft gun mounting in the Sinai desert.*
Below *The struggle against the surrounding Arab states has meant constant vigilance. This soldier in his emplacement on the Golan heights is observing the Syrian positions for any sign of movement.*

the fighting and keeps his headquarters near the front of the advancing column. A reporter who accompanied General Ariel 'Arik' Sharon, the commander of an armoured corps during the Sinai campaign in the Six Day War of 1967, remarked while on the scene that Sharon:

> is not conducting his assault on Egyptian positions by the mechanized and fast-moving army from his headquarters, but surges forward with his units in a command tank . . . he is constantly on the move, issuing orders, listening to messages, working out tactics, sticking his head out to watch the progress of the assault launched under a deafening barrage, and chewing biscuits – the only food he has had in three days. He has had no sleep or rest since he first ordered the armour to press forward.

The inevitable consequence of the 'follow me' principle is a high rate of casualties among the leaders, and in both the 1956 and 1967 campaigns almost half of the Israelis killed in action were officers, including several colonels. This in turn led to a certain amount of disquiet, and a number of Israelis have asked from time to time whether the price was not too high. In reply, the Chief of Staff, General I. Rabin, stated:

> I believe that the fact that our senior commanders go with their men where the danger is, is first and foremost an expression of a certain moral level and quality of humanity . . . our commanders consider it a great personal responsibility to be with their men at the place where the mission is to be carried out, the place where they can have the maximum personal influence on the outcome. I do not know of any single factor to which so much of historic achievement can be attributed as the human and moral quality of our commanders; of which the readiness to go in the van, their personal valour, their audacity, their readiness to risk their lives are direct products. In order to continue to have the kind of army which Israel requires, we need commanders whose quality as men, whose moral level, and whose sense of responsibility dictates that they shall be at the head of their men, in the front line.

The 'kind of army which Israel requires' has been essentially one geared to taking the offensive (although not all Israelis see it this way). As most Arab countries would not acknowledge for many years the Jewish state's right to exist and some vowed to destroy it, some Israelis have argued that any spontaneous Israeli military act, including invasion of neighbouring countries, is a 'preemptive strike' and a defensive manoeuvre; the military advantage gained by such an act was felt to outweigh the effect on world opinion, which generally dislikes the side which shoots first. And up to 1967, of course, the Israeli army had no front on which it had the space to conduct defensive manoeuvres. The so-called Israeli Defence Force, therefore, was a great believer in the principle that the best defence is a good offensive, and its principles for conducting campaigns have been a reflection of this. The combat units do not wait for supplies to catch up, except for fuel and water; they are taught to press forward relentlessly, if necessary without food or sleep, and they are assured (usually correctly) that the enemy will thereby become more exhausted than the Israelis themselves. Night reconnaissance patrols were done away with, being replaced by vanguards under high-ranking officers, which cleared the way for the main force rather than simply collecting information and reporting back, thereby inevitably slowing down the momentum of the thrust.

This offensive spirit and aggressive leadership has only worked, however, because the soldiers themselves have proved themselves capable of sustaining tactics which involve great mobility and the risk of being isolated. As we have seen, reservists undergo regular training and are always competent in the handling of the often new weapons they are called on to use. They had initially a standard of education which was generally higher than that of their opponents. In 1967 this not only made them more effective at handling sophisticated equipment, but also gave small units a sense of responsibility and perspective. Connected to this high standard of education is an intellectual attitude to orders: an Israeli soldier might well refuse to obey orders which clashed with his conscience.

More important than these two elements, however, is the sense of solidarity. In the major campaigns of 1967 and 1973, and in the border strikes and commando-style raids such as the rescue of the hostages at Entebbe, every soldier has known he is fighting for the existence of his homeland and for his people; and he knows why he is fighting, and that this motive is shared by all. The consequence of this is that the main obligation felt by Israeli soldiers is to their comrades-in-arms; it is not to kill Arabs. No wounded soldier is left on the battlefield, whatever the circumstances, even if this means carrying him on a stretcher into the next assault for lack of transport to rear areas. Men even sacrifice themselves trying to recover dead bodies, to prevent mutilation. This attitude has its drawbacks – morale dropped during the Six Day War whenever the wounded could not be flown out by helicopter – but the strong sense of community gives the army an inner strength, a cohesion normally absent in its enemies. This is connected, of course, to the tightly knit nature of Israeli society; the activities of all units of the army can easily be followed in newspapers, and there is an enormous stigma attached to units which perform badly.

Mutual support, the close relationships between officers and other ranks, and an aggressive attitude towards the enemy all reflect the eternal Jewish theme of 'few against many'. If this attitude has often led to their persecution by others in the past, it has also ensured their survival as a people and makes the Israeli soldier one of the most effective fighting men of the modern world. Compromise is still a slow and painful process, and in the meantime Israel remains a 'garrison state' in which all Jewish citizens are dedicated to survival through the ready use of armed force against their enemies.

Tank Warfare in the Desert

Israeli military success against the neighbouring Arab states in the 1956, 1967 and 1973 wars was largely due to the superior fighting qualities of the Israeli soldiers; not so much in terms of personal bravery as in greater technical ability, especially in the use of armoured vehicles.

Fighting in the Sinai, for example, was characterised by the distances and problems of supply. The Israeli solution was the 'Ugdah' formation. Tanks would advance, followed by mechanised troops and infantry. The main supply column would maintain a supply shuttle to the forward units. When enemy resistance was met, the tanks would combine to punch a hole through it along the most important road, and continue their advance. The mechanised units would seal off any dangerous points and continue behind the tanks, while the infantry mopped up. All the while, the shuttle would keep up with the forward units, so that the momentum of the advance could be maintained. The vulnerable main supply column would come through when all resistance had been mopped up.

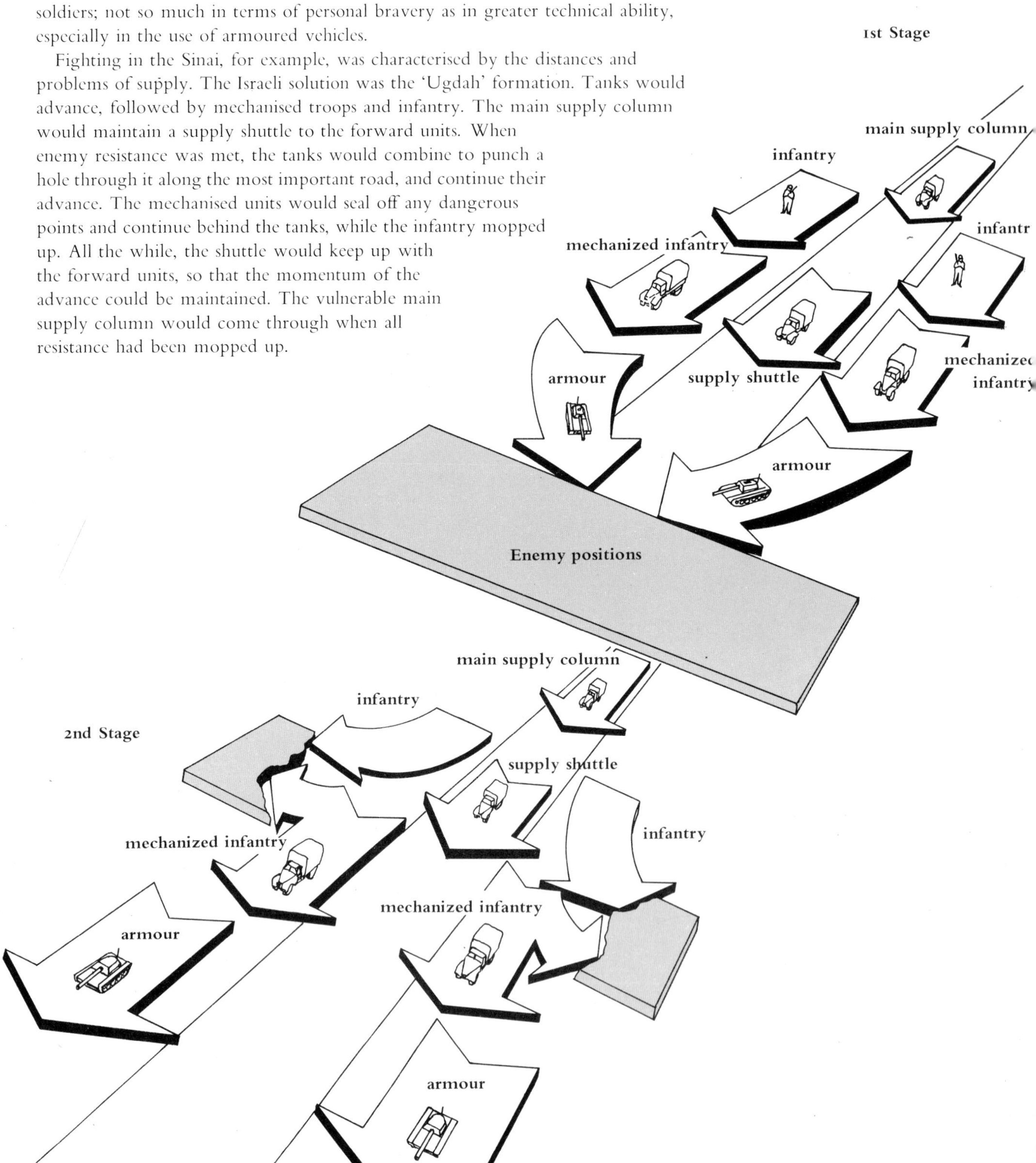

1st Stage

main supply column

infantry

infantry

mechanized infantry

mechanized infantry

armour

supply shuttle

armour

Enemy positions

main supply column

infantry

2nd Stage

supply shuttle

infantry

mechanized infantry

mechanized infantry

armour

armour

armour

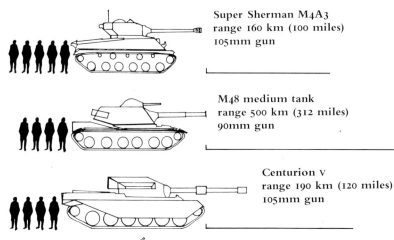

Super Sherman M4A3
range 160 km (100 miles)
105mm gun

M48 medium tank
range 500 km (312 miles)
90mm gun

Centurion V
range 190 km (120 miles)
105mm gun

T54
range 480 km (300 miles)
100mm gun

Three of the most important tanks used by the Israelis were the Super Sherman M4A3, the long ranging M48 and the Centurion; the latter in particular was a fearsome battle tank with a very powerful gun. Its major adversary was the Russian-built T54, reliable and well armoured.

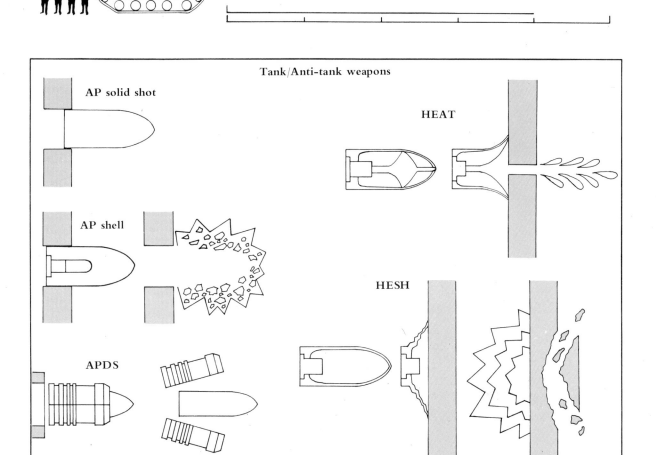

Tank/Anti-tank weapons

AP solid shot

HEAT

AP shell

HESH

APDS

Anti-tank weaponry became steadily more complex during the 1960s and 1970s. The most common types are shown here. Armour Piercing Solid Shot relies on sheer velocity to penetrate armour. Armour Piercing Shell uses velocity and explosive effect as well. APDS (Armoured Piercing Discarding Sabot) is a further refinement, in which part of the shell casing drops away to improve velocity. HEAT (High Explosive Anti-Tank) works on a different principle: the shell has a hollow head which concentrates the effect of the explosive on a narrow point. HESH (High Explosive Squash Head) is a system whereby the shell has a base fuse and thin casing. As it hits the tank, the casing buckles, and the explosive is squashed against the armour before the fuse is activated. The resulting explosion creates shock waves and blasts fragments from the armour into the tank's interior.

INDEX

Page numbers in italics refer to captions

ACKNOWLEDGEMENTS

Title page: Musée de l'Armee/IGDA, page 6: US Marine Corps, 11–12: S. Halliday, 13–22: IGDA, 23: Mansell Collection, 24: IGDA, 25: Alinari, 27: Scala, 31–33: S. Halliday, 34: Metropolitan Museum of Art, 35: IGDA, 40: British Museum, 41: Oldsaksamling, Oslo, 42: S. Halliday, 43: British Museum, 47: Caisse Nationale, Paris, 48–49 M. Holford, 49: Caisse Nationale, Paris, 53: IGDA, 55: Seemüller/IGDA, 56–57: IGDA, 58: Seemüller/IGDA, 59–62: British Library, 63: Dagli Orti/IGDA, 64–65 top: British Library, 65 bottom: Bodleian Library, 67–71: Scala, 74: IGDA, 74–76: Scala, 79: Aargauische Kantonsbibliothek, 81: Kupferstichkabinett, Basle, 81 bottom: Bibliothèque Royale, Brussels, 83: Mansell Collection, 86–87: Mansell Collection, 88–89: Scala, 90–93: Mansell Collection, 94–97: National Army Museum, 100–101: Mansell Collection, 104–106: Bildarchiv Preussischer Kulturbesitz, 107: Robert Hunt Library, 108 left: Victoria & Albert Museum, 108–116: Bildarchiv Preussischer Kulturbesitz, 117: Victoria & Albert Museum, 120: R. Viollet, 121–122: MARS, 123: R. Viollet, 124: Victoria & Albert Museum, 124–125: Giraudon, 127–130: IGDA, 131–133: R. Viollet, 135: Armstrong Roberts/ZEFA, 137: Robert Hunt Library, 138 top: National Library of Medicine, 138 bottom: US National Archives, 140: Armstrong Roberts/ZEFA, 141–143: Robert Hunt Library, 147 top: US National Archives, 147 bottom: Armstrong Roberts/ZEFA, 148: ZEFA, 149: US Signal Corps, 151–152 top: Imperial War Museum, 152–153: Mansell Collection, 154–155: Imperial War Museum, 156–157: Robert Hunt Library, 158: E.C.P.A., 159: Robert Hunt Library: 162: Robert Hunt Library, 166: Orbis Publishing Ltd., 167: Bundesarchiv, 169 top: Orbis Publishing Ltd., 169 bottom: Bundesarchiv, 170 top: Orbis Publishing Ltd., 170 bottom: Imperial War Museum, 171: Bundesarchiv, 174–175 top: Orbis Publishing Ltd., 175 bottom: Bundesarchiv, 174–175 top: Orbis Publishing Ltd., 175 bottom: Bundesarchiv, 179–187: Imperial War Museum, 189: Fujifotos/MARS, 191–195: Robert Hunt Library, 196: Domenica del Corriere, 197–200: Robert Hunt Library, 201–202: US Defense Department, 206–207: Robert Hunt Library, 209: Imperial War Museum, 213: René Dazy, Paris, 216–217: E.C.P.A., 219–221: Keystone Press Agency, 224: E.C.P.A., 225–227: Keystone Press Agency, 229: Topix, 230 top: Private Collection, 230 bottom: John Topham, 232 top: Keystone Press Agency, 232 bottom: Popperfoto.